Springer-Verlag Italia Srl.

Cerebrospinal Fluid Analysis in Multiple Sclerosis

E.J. THOMPSON
Institute of Neurology
University of London
The National Hospital for Neurology & Neurosurgery

M. TROJANO
Istituto di Clinica delle Malattie Nervose e Mentali
Università di Bari
Ospedale Policlinico

P. LIVREA
Istituto di Clinica delle Malattie Nervose e Mentali
Università di Bari
Ospedale Policlinico

Springer

Acknowledgement

The Authors wish to thank FARMADES - Schering Group for the support and help in the realization and promotion of this volume.

ISBN 978-88-470-2207-2 ISBN 978-88-470-2205-8 (eBook)
DOI 10.1007/978-88-470-2205-8

Library of Congress Cataloging-in-Publication Data. Cerebrospinal Fluid Analysis in Multiple Sclerosis / [edited by] E.J. Thompson; M. Trojano; P. Livrea p. cm. Includes bibliographical references and index. 1. Multiple Sclerosis – Pathophysiology. 2. Cerebrospinal fluid – Pathophysiology. 3. Multiple sclerosis – Diagnosis. 4. Cerebrospinal fluid – Examination. I. Thompson, E.J. II. Trojano, M. (Maria), 1952- .III. Livrea, P. (Paolo), 1946- . [DNLM: 1. Multiple Sclerosis – cerebrospinal fluid – congresses. 2. Multiple Sclerosis – immunology – congresses. 3. Cerebrospinal Fluid – immunology – congresses. 4. Cerebrospinal Fluid – chemistry – congresses. 5. Cerebrospinal Fluid Proteins – analysis – congresses. WL 360 C414 1996] RC377.C47 1996 616.8'3407 – dc20 DNLM/DLC for Library of Congress 96-26146

Originally published by Springer-Verlag Italia, Milano in 1996
Softcover reprint of the hardcover 1st edition 1996

Typesetting: Compostudio, Cernusco sul Naviglio (Milano)

Preface

In the spring of 1993 a meeting of the European Charcot Foundation was held in Bari as it was quite clear that there was a renaissance of Italian interest in cerebrospinal fluid. Two of the influential figures with a long-standing commitment to this field were Professor Paolo Livrea and Dr. Maria Trojano. At this meeting we considered the possibility of collecting together the various papers which had been presented, however the idea evolved further to the present volume, which incorporates a broader view. This starts with the anatomical approach, including neuropathology, CSF cells and MRI correlations. It also includes virology, physiology of the blood-CSF barrier and the immunology of intrathecal responses. It further considers the relevance of the CSF parameters to therapy. Finally, the importance of quality assurance and handling of the precious fluid is discussed. The original plan was to have the book published in Italian, but we are grateful to the publishers for allowing English text, in order to have a wider, international audience. As is fate of any book, interest in this area has been accelerated considerably; nevertheless we hope this provides a summary from a multi-disciplinary point of view. From a purely personal point of view, it certainly reflects the warmth of feeling which was evident at the meeting, and which led to the toast "Arrivederci, Bari".

E.J. THOMPSON

Table of Contents

List of Contributors

ANNUNZIATA Pasquale 105
AVOLIO Carlo 79
CAPELLO Elisabetta 15
FERRANTE Pasquale 1
FREQUIN Stephanus T.F.M. 113
GALLO Paolo 93
GIANNINI Paolo 123
GIOVANNONI Gavin 29
HOMMES Otto R. 113
LAMERS Karel J.B. 113
LIVREA Paolo 79, 123, 143
MANCARDI Gianluigi 15
MANCUSO Roberta 1
MARROSU Maria Giovanna 73
ÖHMAN Sten 133
REIBER Hansotto 51
RUGGIERI Maddalena 79
SIMONE Isabella Laura 79, 123
SINDIC Christian J.M. 41
TAVOLATO Bruno 93
THOMPSON Edward J. 29
TORTORELLA Carla 123
TROJANO Maria 79, 123

Cerebrospinal fluid virological analysis in multiple sclerosis

P. Ferrante and R. Mancuso

Viral etiology of MS

Introduction

The detection of virus-specific antibodies, of antigens, and the more recent detection of viral nucleic acids in cerebrospinal fluid (CSF) are powerful tools for the diagnosis of the most common viral diseases affecting the CNS, such as acute and subacute encephalomyelitis and other neurological diseases caused by the replication of viruses in the CNS. However, these virological techniques, including those most recently developed, are not so efficient in cases involving the verification of whether a viral agent may be responsible for such a complex disease as multiple sclerosis (MS).

Since the time of its clinical classification in the 19th century, Charcot himself suggested that there was a possibility that MS was due to infectious agents. Despite all the research that has been conducted and the widening of our knowledge concerning the most varied aspects of the disease, the etiology of MS is still unsolved. In addition, the virus or viruses responsible for the disease have not yet been identified. However, the hypothesis of a viral cause of MS is now an accepted constant in the scientific literatures of the disease [1], regardless of the setbacks encountered by researchers.

A model for the viral etiology

Maybe owing also to its vagueness, the most commonly accepted concept is that MS is a disease caused by an autoimmune process that originates in subjects with genetic predisposition as a consequence of the intervention of exogenous factors and, probably, of one or more viral agents. The formulation of this general hypothesis gives rise to the issue of developing a more involved theory, that is if the aim is to set up virological laboratory studies to verify the viral etiology of MS.

One basic point that must be kept in mind in approaching this issue is the fact that MS is neither an infectious, nor an acute disease, but due to the characteris-

---

Laboratory of Biology, Don C. Gnocchi Foundation IRCCS, Via Capecelatro 66, 20148 Milan, Italy

tics which we have briefly mentioned it is a disease with a natural course resembling that of chronic degenerative diseases. Thus, it is necessary to verify the hypothesis that one or more viruses may represent the etiological agents or, at the very least, co-factors in the etiology of a disease that, while presenting definitive inflammatory phenomena, also presents the typical course of chronic degenerative diseases.

The possible intervening mechanisms of viruses in the development of MS are reported in Fig. 1, as related to the natural course of the disease which, as in the case of chronic degenerative diseases, is divided into five phases: susceptibility, induction, preclinical phase, degenerative phase and rehabilitation. The model taken into consideration in the hypothesis outlined in the figure is that of a persistent viral infection which many human viruses, such as *Herpesvirus*, *Polyomavirus*, *Retrovirus* and occasionally others, are capable of determining.

Albeit indirectly, the model of persistent viral infection already supplies an answer to one basic question: why are viruses, and not other known micro-organisms, suspected in the etiology of MS? In fact, it is evident that the persistent viruses, which not only necessarily replicate within cells, but persist in the cytoplasm or nuclei and in some cases, like *Retroviridae*, even become part of the cellular genome, present ideal characteristics as candidates for etiological agents of MS.

Some aspects regarding the hypothesis outlined in Fig. 1 require further discussion, particularly the proposed intervening mechanisms of the viruses in the natural course of MS. It is known that susceptibility to the disease is, to some extent, genetically determined and that, in a similar manner, the viral infections may also follow different courses depending on the genetic characteristics of the infected host. As shown by epidemiological studies, the period of susceptibility to MS begins in early infancy and ends in adulthood. In keeping with the postulated viral etiology, the action of a virus during the susceptibility period leads to the induction phase, which is

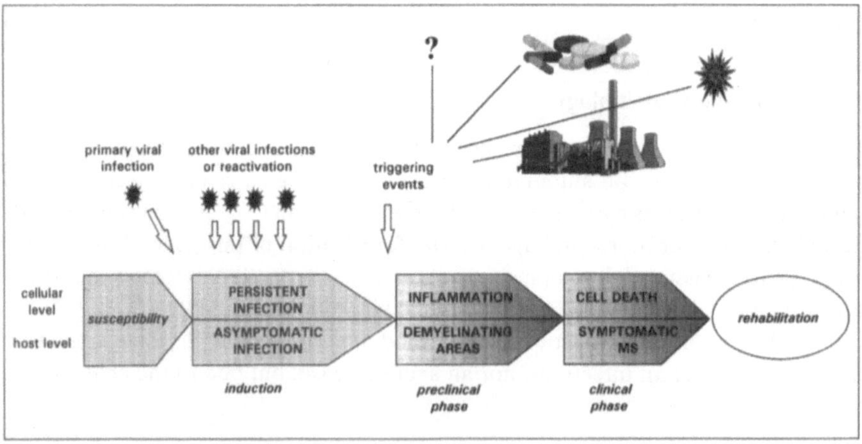

Fig. 1. A model representing the possible action of a virus-induced persistent infection on the natural course of multiple sclerosis

an unidentifiable point in time probably because it coincides with cell or tissue alterations not perceptible on clinical examination or testing. In addition, induction is a process that requires additional time before onset of the disease takes place, that is the transition to the preclinical phase.

It is also known that many viruses capable of inducing chronic infection, such as the previously mentioned *Herpesvirus* and *Polyomavirus*, as well as many other viral agents, give rise to ubiquitous infections that affect most subjects during childhood in a symptomatic or aysmptomatic manner. It is postulated that the first event in the induction of the disease takes place at the time of the primary infection by one of these viruses, in subjects who are susceptible because of their genetical disposition, and this event may be simultaneous with the triggering of the autoimmune mechanism leading to the disease. Afterwards, with each reactivation of the virus, a strengthening of the autoimmune action may take place, even in cases in which the viral reactivation does not give rise to clinical evidence and even in the presence of a partial reactivation, with incomplete transcription and translation of a limited amount of specific proteins.

The transition to the preclinical phase is thought to take place as a consequence of the intervention of another factor, which may consist in infection by another virus or, once again, in a reactivation of the same virus. It can be reasonably postulated that any infection, even by non-viral agents, or other events such as trauma, stress, hormonal disorders, chemical exposure, pharmacological treatment, etc., can function as triggering factors of MS in subjects in whom a pathogenic mechanism has already been induced. It should also be further specified that even the start of the preclinical phase is difficult to determine; it is probable that a clear-cut borderline between the induction and the preclinical phases of MS does not exist. Yet, the preclinical phase is of considerable importance as a target of laboratory testing, since the development of techniques permitting the detection of lesions when clinical signs of MS are still not evident will lend greater weight to laboratory investigations in terms of our understanding of MS etiology. Moreover, besides its particular significance for the study of MS etiology, the study of infection, reactivation or viral replication phenomena occurring in the preclinical phase may also lead to the pinpointing of targets for specific therapeutic approaches.

Verification of the hypothesis

The hypothesis of a viral etiology of MS is therefore a complex one, and its verification requires the use of the complete array of the most sophisticated investigative tools available for virological analyses, to be used with the most suitable samples collected in the most significant phases of the disease.

Which samples are most suited to verifying the viral etiology of MS? One of the most spontaneous answers to this question is: brain tissue samples. Considering the pathogenesis of MS, and that the brain is the site of the lesions determining the disease, it thus represents the ideal target for virological screening. However, this is not entirely true for several reasons which we shall briefly consider. In the first place, it should be kept in mind that the collection of samples using invasive techniques like stereotactic biopsy cannot be considered in the case of MS patients for ethical reasons, and also because this technique does not permit the collection of targeted specimens as precisely as needed. Thus, brain tissues can only be collected upon autopsy, but also in this case the specimen bears the risk of being of little consequence, considering the long period of time that fortunately passes between disease onset and patient's death. Working with autopsy specimens, it is possible to detect the presence of any viruses directly in the demyelinating areas, and also to distinguish between active and older plaques.

On one hand, the failure to detect a virus in autoptic brain tissue may also be attributable to a clearance of the virus over time. On the other hand, the detection of a virus, viral nucleic acid or viral antigens, which in the demyelinating areas may not be of etiological significance, could be due instead, to the fact that the virus was localized in the CNS much earlier. Some indirect evidence suggesting this possibility consists in the ever more frequent observation of viruses, and particularly of viral nucleic acids, in brain tissues collected at autopsy from subjects whose deaths were due to causes other than CNS infection [2]. Also the detection of viral mRNA or proteins, clearly indicating viral replication, may not necessarily bear the significance of a resolving factor, since it may be due to the indirect effects of the activation of inflammatory cells, which are present in demyelinating areas.

On the contrary, CSF proves itself to be an ideal specimen for studies investigating the viral etiology of MS. The collection of CSF samples is not as traumatic for the patient as a biopsy, and at the same time the elements observed in the CSF are sufficiently indicative of what is happening in the CNS. CSF samples are often collected form MS patients to perform laboratory investigations which include the search for IgG oligoclonal bands (OB) and the measurement of intrathecal synthesis indexes which are useful for the diagnosis of MS. CSF is thus collected at a very early stage of the clinical phase and, as above mentioned, this lends greater weight to the results in terms of their etiological significance. Moreover, it is also important to note that it is possible to detect viruses in CSF by direct methods such as antigen detection, nucleic acid hybridization, electron microscopy, or by means of culture isolation assays, and it is also possible to search for indirect evidence of a viral infection of the CNS by measuring the viral antibodies and screening for viral-reactive cell clones.

What virological methods are currently available? Virological laboratory techniques essentially consist of direct methods of searching for viruses or their antigens, in the isolation of the viruses by cultivation and in viral antibody testing. All these methods have been used for MS studies, often with conflicting results. Yet, the constant process of perfecting virological techniques still makes their application necessary in the attempt to clarify the etiology of MS. The main aspects involved in the utilization of virological techniques on CSF in MS will be evaluated in the following paragraphs.

CSF viral antibodies

Background

Great encouragement for virological studies in MS came from the report of Adams and Imagawe [3], who published in 1962 the finding of increased levels of measles antibodies in MS patients. These observations, followed by many other reports [4], strongly influenced the virological research on MS also in consideration of the pathogenetic characteristics of measles virus.

Measles virus is well known for its capability of inducing Subacute Sclerosing Panencephalitis (SSPE), a rare fatal disease of CNS, and shows some epidemiological features that, according to many Authors, could well fit with the ideal characteristics of the putative *MS virus*. In particular, measles induces a common exanthematic disease which occurs early in childhood in the southern countries of the northern hemisphere, and later in the temperate and cold regions, with a gradient that is quite similar to that of MS distribution, whose prevalence is higher in the north.

Many other viruses have been studied by looking for specific antibodies in serum and CSF and comparing the levels found in MS patients with those of control groups (Table 1). Both DNA and RNA employed viruses including *orthomyxovirus, paramyxovirus, vaccinia virus, rubella virus, human herpesvirus, retrovirus* and others have been investigated on the basis of epidemiological considerations and because of the immunological similarities between viral envelopes and cell membranes. Moreover, specific antibodies toward naked virus such as *poliovirus* have been also looked for in MS patients.

On the whole, the large amount of data obtained with the viral antibody methods have often been controversial and did not indicate the solution of the MS etiology enigma. This failure is probably due to the fact that the difference in the viral antibody levels observed by comparing MS patients and controls could be due to the immunological abnormalities that characterize MS.

Table 1. Some of the viruses for which high antibody levels have been detected in MS patients

- Measles virus
- Canine distemper virus
- Mumps virus
- Parainfluenza viruses type 1 and 3
- Rubella virus
- Influenza viruses type A, B and C
- Herpes simplex virus type 1 and 2 (HSV 1 and 2)
- Varicella Zoster virus (VZV)
- Epstein Barr virus (FBV)
- Human Cytomegalovirus (HCMV)
- Poliovirus
- Coronavirus
- Vaccinia virus
- Human T lymphotropic virus type I (HTLVI)
- Paramyxovirus SV5
- Human endogenous retrovirus (HRES-1)

In other words, the autoimmune phenomena and the status of chronic activation of the immune system could be the cause of the elevated CSF antibody levels often observed in MS patients toward many viral agents. This could be true in particular when CSF antibodies are looked for toward viruses which induce common infection and, thus, a large majority of the general population in the western countries have circulating specific antibodies. On the contrary, when exotic viruses (i.e. *ebola virus*) or rare viruses (i.e. *retrovirus* or animal viruses) that usually do nor infect man (i.e. *canine distemper virus* or *rabies* or *borna virus*) are taken into consideration, the finding of CSF antibodies is more significant for the MS etiology.

On this basis it can be accepted that in MS studies the viral antibody searches are unable to confirm or reject the viral etiology, but are useful as generators of new etiological hypotheses.

Technical aspects

The methods for the search of viral antibodies, usually called serological methods, are numerous and have different properties and limits. In Table 2 the most common serological tests and some of their characteristics are reported.

The search for viral antibodies in paired samples of serum and CSF is a classical method for diagnosing CNS viral infection, since the detection of antibodies produced within the CNS is one of the most important parameters for confirmation of a neurological viral infection. The intrathecal synthesis of specific viral antibodies is the result of their production and secretion from the B-cells that enter together with other immune cells into the CNS during viral infection. A critical point is the fact that, like other molecules, immunoglobulins can also diffuse from the blood into the CSF when their levels are particularly high in the serum. Thus, during a systemic, non-CNS infection, viral antibodies could be found in the CSF as a consequence of this non-specific diffusion. However, on this occasion, because of the blood-barrier, viral antibodies will be present in a lower concentration in CSF compared to serum. For a long time it has been accepted that antibody levels to a certain virus are usually about 1:400 times lower in CSF than those in the serum [4], and that a significant increase in that ratio indicates intrathecal synthesis probably stimulated by a CNS viral infection. This is a very simple approach and can be efficient for the diagnosis of acute viral infection of CNS but, of course, it is less useful when applyed to subacute or chronic diseases such as MS.

The most widely used method for determining intrathecal antibody synthesis is to evaluate IgG index [5]. The significance and the various aspects of this and of other indexes are discussed elsewere in this book. With regard to the viral antibodies, it must be remembered that the serum/CSF ratio of specific antibody can be adjusted for total IgG content to correct for possible increased non-specific permeability of the blood-brain barrier using the following formula [6]:

$$\text{Virus-Specific Antibody Index: } \frac{\text{virus-specific IgG in CSF/total IgG in CSF}}{\text{virus specific IgG in serum /total IgG in serum}}$$

Table 2. Some of the most common methods for viral antibody search and their principal characteristics

Complement Fixation (CF) CF is a classical virological method often with specificity and sensitivity problems. It can be used for antibody titration by serial dilution. CF antibodies are usually toward internal viral antigens. CF can be only partially automated.
Hemagglutination Inhibition (HI) HI has the same characteristics of the CF test; however, it detects antibodies against viral surface proteins.
Neutralization Assay (NA) NA shows many common characteristics with HI, but NA is more difficult to execute and is time-consuming since it must be performed in cell culture.
Enzyme Linked Immunosorbent Assay (ELISA) ELISAs are a wide family of methods for antibody testing. They are sensitive, specific and can be easily automated. ELISA can contain antigens viral lysate, recombined proteins and viral peptides.
Radio Immuno Assay (RIA) RIA has the same advantage and flexibility of ELISA and perhaps it is more sensitive. Because radiolabeled products are used, RIA can be performed only in special laboratories and its use should be limited.
Immuno Fluorescent Assay (IFA) IFA is a low-cost, easy to perform antibody test. It cannot be automated and has sensitivity and specificity lower than ELISA and RIA.
Western Blot (WB) Sensitive and specific, WB gives the possibility of analyzing antibody toward each separate viral antigen. It cannot be employed for quantitative analysis of antibody response.
Radio Immuno Precipitation Assay (RIPA) RIPA has the same properties of WB but is more difficult to perform and time-consuming; moreover, it uses radiolabeled aminoacids and therefore requires special facilities.
Recombinant Immuno Blot Assay (RIBA) This method is performed as a WB after a previous mechanical deposition of antigens on the nitrocellulose membrane.
Iso Electro Focusing (IEF) with Immunoblot This is a new method that can be used to assess antigen specificity of IgG OB found in CSF. The transfer of OB after IEF to a nitrocellulose membrane previously impregnated with the relevant viral antigen can be a strong indicator of the involvement of this virus in the CNS disease.

A virus-specific antibody index of >1 can be considered strong evidence for intrathecal synthesis and thus for CNS infection.

Among the other formulae that have heen proposed to define the specific intrathecal antibody must be remembered the Antibody-Index which has been defined by Reiber and Lange [7].

The CSF MRZ reaction

As previously discussed, the finding of viral antibodies in the CSF seems not to be relevant for the etiology of MS. The presence of intrathecal viral antibody synthesis can be observed during an acute CNS infection with reaction to viral antigens, but of course this is not the case in MS, which has a chronic course. The presence of high viral antibody levels within the CSF, often found in MS patients, could be explained as a part of a polyspecific immune response that is typical for chronic inflammatory diseases. For these reasons it has been suggested that the measurement of the antiviral response againts measles, rubella and varicella-zoster virus (MRZ reaction) in CSF and matched serum samples of MS patients could be employed as a paraclinical test for the diagnosis. The MRZ reaction is conducted by measuring the antibody levels toward the three viruses with ELISA test, that must be standardized in order to obtain accurate quantitative data [8]. In this way the Antibody Specific Index (ASI) can be defined (see Chap. 3 of this book).

The application of the MRZ reaction in the diagnosis of MS seems promising so far, but more studies are needed in order to assess the diagnostic significance and the possible prognostic relevance of this test.

CSF virus isolation

Virus isolation is one of the oldest and most common methods employed for the diagnosis of viral infections. The isolation and the identification of a virus from CSF can be considered as the gold standard for diagnosing viral infection of the CNS. Since viruses replicate exclusively in living cells, virus isolation must be performed on cellular substrates that in virological laboratories include laboratory animals, embryonated eggs and cell cultures. Thus, is order to isolate viruses, the laboratory will mantain all these living substrates. Among the cell lines a large variety of cellular types exist, including human diploid cell lines, primary cell lines, transformed cell lines and lymphoblastic cell lines.

In the course of CNS viral infection, many different samples including throat swabs, blood, urine, faeces and CSF must be collected according to the replication sites of the suspected virus. While all the other specimens need preliminary treatment to avoid bacterial contamination or toxic effect on the cell culture, CSF can be used without any preliminary treatment. Sometimes CSF shows a toxic effect on some cell lines and in this case 12 or 24 hours after the inoculum the supernatant can be removed, assuming that if a virus is present in CSF this time is sufficient for its penetration into susceptible cells.

Virus isolation is a very sensitive method since it can permit viral replication even if a very small number of viral particles is present in the sample.

The great limitation of viral isolation derives from the fact that a large number of viruses do not replicate on the cell lines that are currently available, and therefore this method can give false negative results. Over the years many attempts at virus isolation have also been performed on CSF and other samples collected from MS patients. Many conventional viruses or conventional infectious agents have

Table 3. Some viruses that have been isolated from samples, including CSF, collected from MS patients. The year in which the virus was isolated is indicated in parentheses

- Rhabdovirus (1946)
- Herpes simplex virus type 2 (HSV2) (1964)
- Scrapie agent (1966)
- Multiple Sclerosis Associated Agent (MSAA) (1972)
- Parainfluenza virus type 1 (1972)
- Measles virus (1976)
- Bone Marrow Agent (1978)
- Coronavirus (1979)
- Cytomegalovirus (CMV) (1979)
- Retrovirus-like agent (LM7) (1989)

been isolated from various laboratories; the most relevant of these observations are reported in Table 3.

Unfortunately, after the first report many of these results were not confirmed in other laboratories. The last relevant observation of a viral agent in MS samples is that of Perron et al. [9] who isolated a retrovirus-like agent. This finding gave new impetus to the study of the possible involvement of retroviruses in MS, that was already suggested when HTLV RNA and antibodies were found in MS patients.

Molecular studies using CSF

Historical background

In recent years there have been far-reaching developments with techniques used in the field of molecular biology and manifold application of these techniques to diagnostic approaches to viral infections. The application of these methods was also further extended to the study of diseases, among which was MS, in which a viral etiology was suspected, with the aim of acquiring evidence to support the hypothesis.

The techniques that may be adopted for research on MS are essentially represented by the hybridization of nucleic acids. Unlike that which takes place in the case of viral isolation or serological assays, hybridization makes it possible to detect a viral infection even when the virus is in a latent state. Hybridization between a labeled DNA or RNA probe and a viral genome (DNA or RNA) can be performed directly on the infected cells or tissues using *in situ* hybridization (ISH). If hybridization is conducted on previously extracted DNA or RNA, or on DNA or RNA extracted from CSF, the technique is instead termed Southern or Northern Blot, respectively.

ISH is a very sensitive and specific technique, and it represents a valid tool for the study of CNS infections. However, the methodology of this technique is rather complex: with ISH, which keeps the structure of the cells and tissues examined, it is possible to detect the cells infected by the virus by using double staining. The Southern and Northern blot techniques are well known and have been widely used for many years in genetic studies. Their employment in the field of virology has proved to be less efficient, as the amount of viral nucleic acids is quite low in biological samples.

Undoubtedly, the most suitable technique to detecte viral genomes in CSF is the polymerase chain reaction (PCR) [10]. PCR analysis permits one to amplify a known genome sequence and obtain millions of copies, thus making it a highly sensitive method. Amplified DNA can then be subjected to hybridization with a probe to verify its specificity, using Southern blotting or liquid hybridization, which can be performed with radioisotope-labeled or enzyme-linked probes.

Using the PCR, it is also possible to amplify viral RNA because a preliminary retrotranscription step can be performed to synthesize the complementary DNA (cDNA) before running the PCR (RT-PCR). To improve the sensitivity of PCR, many laboratories are successfully using the nested PCR (n-PCR), which consists of a preliminary amplification of an external DNA region, followed by a second series of amplification cycles performed with primers which anneal to an internal part of any DNA amplified in the first cycle (Fig. 2).

The extensive application of the n-PCR to the diagnosis of viral infection has also eliminated the uncertainties existing until a few years ago concerning the low specificity of this method.

Technical aspects of PCR analysis of CSF

CSF specimens are excellent candidates for PCR analysis. CSF can be used in its current state without requiring treatment for DNA extraction, whereas brain tissue and peripheral blood samples require preliminary treatment prior to testing.

Our experience in screening CSF samples for viral nucleic acids in the case of samples collected from patients with *Herpes, Human Cytomegalovirus* and Epstein Barr virus encephalitis and progressive multifocal leukoencephalopathy (PML), has led to the establishment of some conditions for running PCRs on CSF. CSF samples are collected, immediately stored on ice, and then examined by PCR; if the PCR is not scheduled to be performed immediately or within a few hours, the CSF can be efficiently stored at –20°C or –80°C temperature, with best results at the colder temperature. Twenty microlitres of CSF are added to a final volume of the PCR reaction mixture, without requiring any treatment. Due to the presumably small amount of viral nucleic acids present in CSF, the n-PCR is preferred to the single PCR. Special care must be taken in cases of viral encephalitis of various types, also during sample collection. In fact, when there is an acute viral infection involving the CNS, viral nucleic acids are detectable in the first few days of the disease; but when virus-specific antibodies are produced, the PCR may yield negative results, probably as a consequence of viral clearance. This phenomenon is also observable when screening CSF for viral antigens to establish a diagnosis of viral encephalitis. Thus, a PCR performed on CSF can yield false-negative results even in the case of acute viral encephalitis if the sample is collected at the wrong point in time. Recent findings concerning Herpes encephalitis support this possibility, as patients with nega-

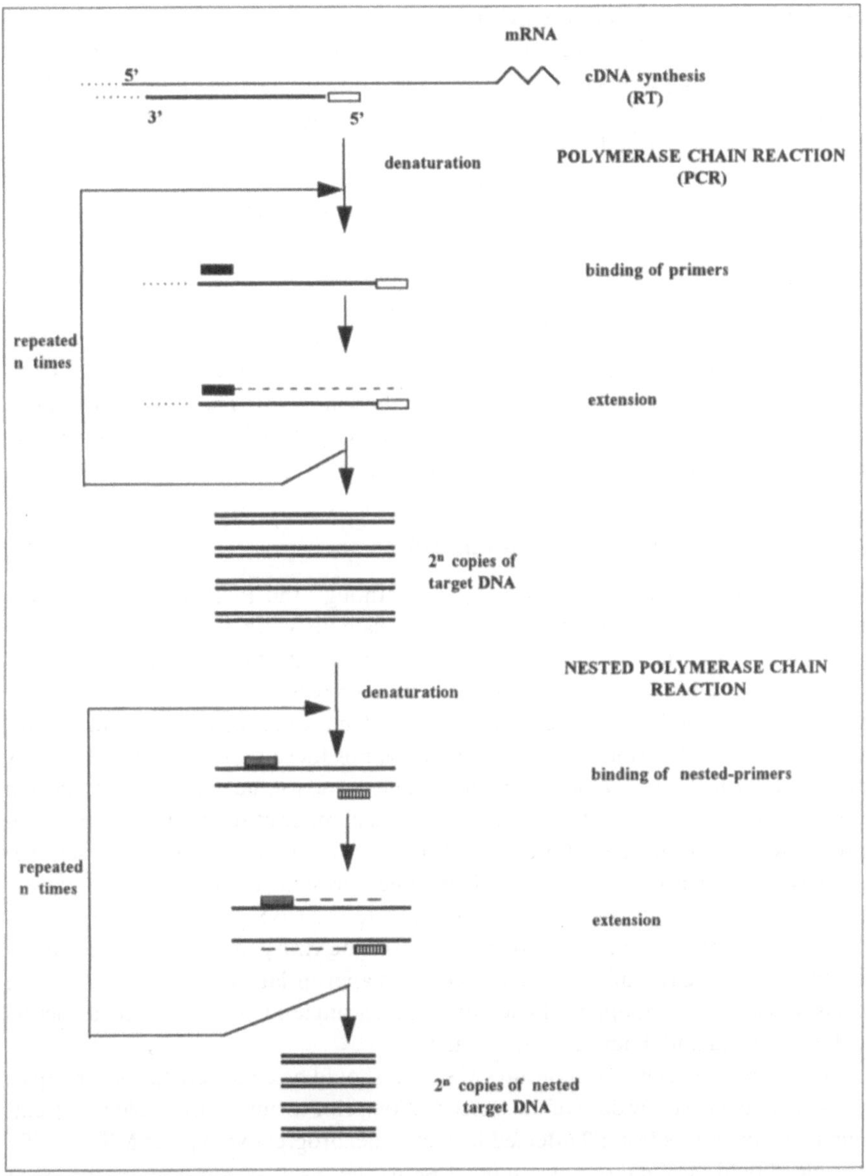

Fig. 2. Outline of the nested polymerase chain reaction analysis, including a preliminary retrotranscription step

tive PCR results for Herpesvirus in CSF showed good clinical responses to treatment with acyclovir [11].

Another aspect to be taken into account is that, as a consequence of an increased permeability of the blood-brain barrier during a systemic viral infection, there is a possibility of the virus being carried from the blood to the CSF. Such an event must be considered as extremely rare; however, some methodological improvements can be adopted to avoid false-positive PCR results. For example, using the same PCR the presence of the virus in the blood should be verified, and if this is the case a quantitative PCR, or, better, a sequence analysis of the amplified products should be performed.

The use of the PCR following a preliminary retrotranscription step, makes it possible not only to amplify genomic RNA, but also to search for messenger RNA, thereby permitting verification of whether the virus is latent or actively replicating. This approach is of particular importance in the study of brain tissue, but it should also be performed when CSF is being examined.

The evaluation of PCR results should take into account the status of the blood-brain-barrier through the assessment of serum and CSF albumin and IgG levels, and the appropriate formula should be applied thereafter [5]. Therefore, the routine performance of CSF analyses at the time of the PCR investigations represents a reliable and important guideline to be followed.

PCR analysis of CSF from subjects with MS

In the previous section we observed that although the PCR is a very efficient method and CSF is particularly indicated as a sample for PCR testing a number of problems exist in the case of diagnosis of acute viral infection of the CNS. The difficulties involved in the detection of any virus in the CSF from MS patients, are obviously augmented by all of the factors discussed in the section on the natural history of MS and the possibility of a viral etiology. For these reasons, though the PCR analysis appears to be a rather simple technique to use for the detection of viruses in CSF samples from MS patients, it does require a number of steps and a complex approach within a well-defined protocol (Table 4). Taking into account that samples are screened for a virus that has probably been present for years in the organism and has then spread to various tissues besided to the CNS, the scheduling of CSF and other sample collection, the identification of the viral genomic regions to be targeted with the PCR, and thus the choice of an appropriate set of primers, and the decision whether to adopt a n-PCR rather than a single PCR, are all critical factors to be evaluated and determined in advance.

In the case of relapsing-remitting MS, CSF should be collected during a relapse, possibly in the first few days of the exacerbation prior to any pharmacological treatment. If a patient is instead affected by a chronic, progressive type of MS, the CSF sample should be analyzed at a time close to the onset of the disease, the point when it is usually collected for MS diagnosis.

Table 4. Guidelines for the employment of PCR in CSF analysis

- Collect the CSF during an acute relapse and in early stage of the disease
- Collect on the same day samples of peripheral blood, urine and saliva
- Perform on the specimens n-PCR tests for more than one virus
- Employ multiple n-PCR in order to amplify different regions of the viral genome

It is of fundamental importance to collect the other types of sample from the patient on the same day. Many different samples may be collected for PCR analysis, depending upon the replication sites of the various viruses. However, sample collection should include at least peripheral blood, urine and saliva, because most viruses that are capable of inducing persistent infections in humans, usually can be detected in one or more of these body fluids. The PCR analysis should be performed on all the samples in order to ascertain whether the positive results obtained with the CSF represent an indicator of viral replication restricted to the CNS or also involving other areas of the body. CSF is a precious material and it should be screened by PCR for more than one virus whenever specimens from MS patients are involved.

Another critical factor in the use of PCR analysis of CSF from MS patients concerns the choice of primers, and thus the part of the genome to be amplified. Usually a highly conserved region of the viral genome is the best target for PCR analysis; nonetheless, it may be important to attempt to amplify sequences belonging to different genes such as the regulatory genes, capsid protein coding genes and, when present, the envelope glycoprotein genes. This approach should prevent false-negative results, that may occur in the case of viral genomic mutations, a relatively frequent event, which for some viruses like JC virus in PML, seems to be important for the development of the neurovirulence [11].

Conclusions

Virological investigations conducted on CSF and other specimens from MS patients have been performed over the past fifty years, but the viral etiology of the disease remains a hypothesis without grounds for confirmation or rejection. Virology, and particularly neurovirology, are rapidly developing fields. The past decade has seen the discovery of many new human viruses worldwide which are capable of inducing persistent infection affecting also the CNS. Virological techniques have seen dramatic improvements: viral antibody assays are now extremely sensitive and specific, and can be applied to a large number of viruses. They can also be performed in almost completely automated systems requiring small amounts of sample material. Moreover, an incredibly large variety of cell lines have become available for isolation assays, providing the possibility of isolating viruses, such as the new human herpes viruses which were unknown until just a few years ago. The application of the methods used in molecular biology to neurovirology is becoming more and more common and it is producing successful results.

The study of CSF is an excellent tool for our understanding of the pathogenesis of MS. The virological analysis of CSF from MS patients should always include the measures described in the approach outlined in the sections above. The virological analysis of CSF, and particularly antibody titers and MRZ reaction, may also represent supplementary diagnostic and predictive tests that could contribute to the establishment of new treatment or to the screening of patients for clinical trials. Our efforts should concentrate on conducting extensive studies which include antibody assays, isolation assays and PCR analysis. Extensive, comprehensive virological studies are expensive and time-consuming. However, when CSF and the other specimens are collected, treated and appropriately stored, they could be analyzed by a network of virology laboratories which would apply their own specific skills in a joint effort to define the viral etiology of MS.

References

1. Kurtzke JF (1993) Epidemiologic evidence for Multiple Sclerosis as an infection. Clin Microb Rev 6: 382-427
2. Ferrante P, Caldarelli-Stefano R, Omodeo-Zorini E, Vago L, Boldorini R, Costanzi G (1995) PCR detection of JC virus DNA in brain tissue from patients with and without Progressive Multifocal Leukoencephalopathy. J Med Virol 47: 219-225
3. Adams JM, Imagawa DT (1962) Measles antibodies in multiple sclerosis. Proc Soc Exp Biol Med 3: 562-566
4. Norrby E (1978) Viral antibodies in multiple sclerosis. Prog Med Virol 24: 1-39
5. Link H, Tibbing G (1977) Principles of albumin and IgG analysis in neurological disorders. Evaluation of IgG synthesis within the central nervous system in multiple sclerosis. Scand J Clin 37: 397-401
6. Salmi A, Reuanen M, Ilonen J, Panelius M (1983) Intrathecal antibody synthesis to virus antigen in multiple sclerosis. Clin Exp Immunol 52: 241-249
7. Reiber H, Lange P (1981) Quantification of virus specific antibodies in cerebrospinal fluid and serum: sensitive and specific detection of antibody synthesis in brain. Clin Chem 37: 1153-1160
8. Felgenhauer K, Schädlich HJ, Nekic M, Ackermann R (1985) Cerebrospinal fluid virus antibodies. A diagnostic indicator for multiple sclerosis? J Neurol Sci 71: 291-299
9. Perron H, Geny C, Laurent A, Mouriquand C, Pellat J, Perret J, Seigneurin JM (1989) Leptomeningeal cell line from multiple sclerosis with reverse transcriptase activity and viral particles. Res Virol 140: 551-561
10. Saiki RK, Sharf S, Faloona F, Mullis KB, Horn GT, Erlich HA and Arnheim N (1985) Enzymatic amplification of β-globin genomic sequences and restriction site analysis for diagnosis of sickle cell anemia. Science 230: 1350
11. Thompson EJ (1995) Cerebrospinal fluid. J Neurol Neurosurg Psychiatry 59: 349-357
12. Martin JD, King DM, Slauch JM (1985) Differences in regulatory sequences of naturally occurring JC virus variants. J Virol 53 (1): 306-311

Neuropathology of multiple sclerosis lesions relevant to cerebrospinal fluid abnormalities

G.L. Mancardi and E. Capello

Introduction

Multiple Sclerosis (MS) is the commonest organic disease of the Central Nervous System (CNS) in the young adult throughout the temperate zone. The hallmark is the demyelinating plaque, in which the removal of apparently normal myelin from axons, usually against a background of perivenular infiltration by lymphocytes, plasma cells and mononuclear cells, is observed.

Immune mediated mechanisms are considered to play a relevant role but the precise immunopathogenesis of MS is far from clear.

An important support for the possible immunological origin of the disorder comes from the presence of immunoglobulins (Ig) and lymphocytes in the cerebrospinal fluid (CSF), the closest structure to CNS parenchyma. The increased local levels of Ig represent the most frequent immunological change in MS and this has practical diagnostic importance. MS is not, however, only related to antibody reactivity toward myelin components as the oligoclonal bands in the CSF might suggest. Strong indication in favour of cellular immunity as pathogenetically significant results from morphological and immunopathological studies of CNS tissue in animal models of MS, experimental allergic encephalomyelitis (EAE) and *in vitro* studies. The humoral and cellular immune changes that can be detected in the CSF are strictly related to the process of chronic inflammation which develops in the CNS.

Gross pathology

At external inspection of the brain no abnormalities are usually detected, apart from a slight degree of cerebral atrophy and widening of the cerebral sulci. In some cases of long duration, firm grey lesions can be directly observed on the ventral surface of the pons and medulla and along the spinal cord, which appears, mainly in the cervical region, shrunken and atrophic. On the cut surface old plaques appear as firm, defined grey areas, mainly distributed around the lateral ventricles

---

Department of Neurological Sciences, University of Genoa, Via De Toni 5, 16132 Genoa, Italy

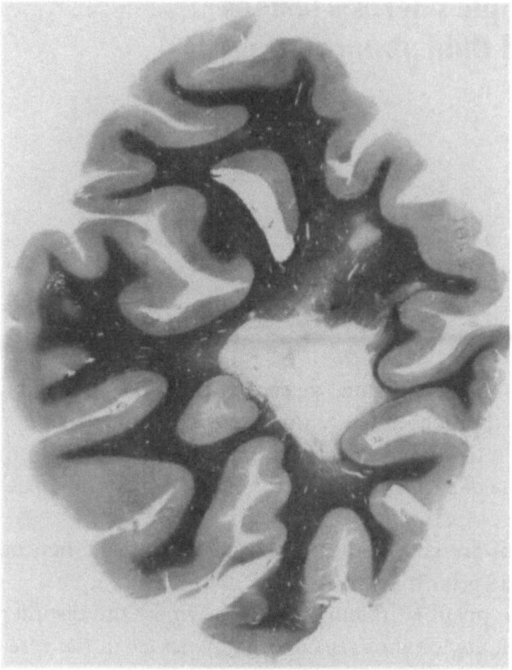

Fig. 1. Periventricular plaque surrounding the posterior horn of the lateral ventricle

(Fig. 1) and around the IV ventricle, but occurring anywhere in the CNS, especially in the central white matter, in the callosal radiations, in the corpus callosum, in the subcortical white matter and in the ventral and dorsal areas of midbrain, pons, medulla and spinal cord. Plaques of recent origin, containing a large amount of macrophages, appear as limited lesions of pink or chalky appearance.

The periventricular location of the plaque is an almost constant finding, and this particular distribution has been attributed to various factors, such as 1) the presence of an unidentified "toxic substance" in the CSF; 2) the abundant venous network of collecting veins in the subependymal area, which drains into the great vein of Galen and could be damaged by back pressure reflux from the internal jugular vein [1]; 3) the presence in this area of several phagocytic cells bearing high avidity IgG Fc receptors, which may behave as antigen-presenting cells [2].

Neuropathology of MS lesions

In a longstanding disorder such as MS, it is usual to find lesions of different histological age, and in the same case old inactive lesions together with chronic plaques with inflammatory edge activity are encountered. On the basis of the extent of inflammatory activity, MS plaques are classified as acute/recent, chronic, chronic active and shadow plaques. It is difficult to establish the initial neuropathological change of the plaque, and different viewpoints exist on the earliest abnormalities that can be detected. Some Authors emphasize the presence of myelin disintegration without increased cellularity [3]; others give priority to the appearance of microglial nodules, which express class II histocompatibility antigens in apparently normal white matter [4], while others consider the perivascular inflammation as the primary lesion. Blood brain barrier (BBB) damage with initial invasion by lymphocytes, which precedes the appearance of macrophages and the event of demyelination, is now considered the primary event [5].

The acute/recent plaque

The acute/recent plaque is markedly hypercellular, due to the presence of perivascular cuffs of inflammatory cells and to the diffuse infiltration of foamy macrophages throughout the lesion. Myelin sheaths disintegrate, and sudanophilic breakdown products are observed within the macrophages, while axons are usually preserved.

The increased cellularity is due: a) to the presence of lymphocytes, mononuclear cells and plasma cells, usually distributed in a perivascular fashion within the Virchow Robin space (Fig. 2), but spreading also in the surrounding parenchyma; and b) to the infiltrating mononuclear cells, macrophages and resident microglial cells, which appear in close contact with fragmented and disrupted thin myelin. The astroglial reaction is usually present, but intense astrogliosis is not a prominent feature. At the centre of the plaque small venules with perivascular inflammation are usually observed, and the inflammation and the demyelinating process usually extend outwards along the course of the veins. Due to the frequent presence of

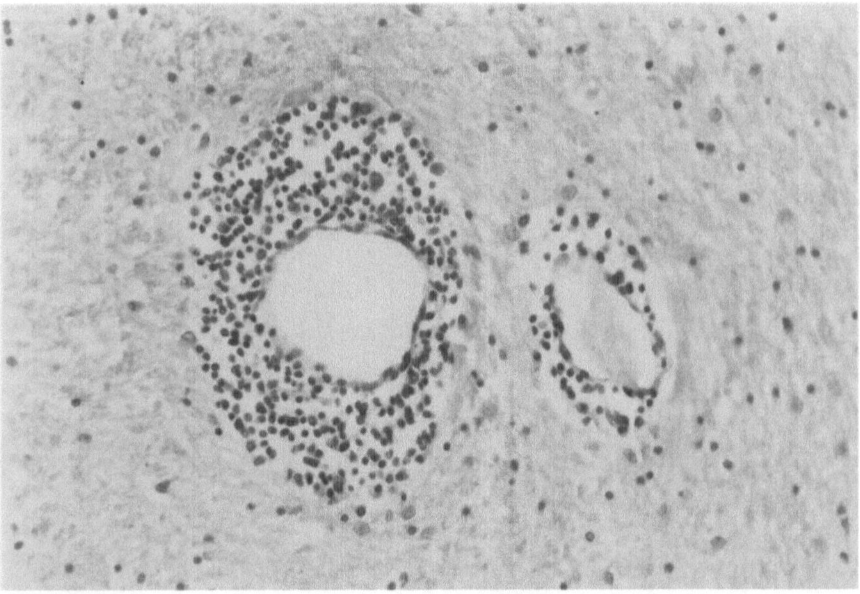

Fig. 2. Perivascular inflammation containing many lymphocytes, mononuclear cells and a few plasma cells

plaques in the ventricular areas, it appears possible that the disease starts near the ventricular surface, around the draining subependymal veins, and then spreads into the brain following the course of veins (Dawson's fingers).

However, though more frequent in the subependymal region, new plaques may arise anywhere in the white matter of the CNS. The damage to the BBB is an early event, characterized by perivascular inflammation, perivenous fibrin and complement deposition and disruption and duplication of vessel basement membrane.

The chronic plaque

The chronic plaques are the most frequent lesions observed in MS brains and are well represented by the large periventricular plaques, always present in MS cases of long duration. Chronic plaques are sharply demarcated from the surrounding white matter and they are usually hypocellular, with complete disappearance of oligodendrocytes in the center of the lesion, although some surviving oligodendrocytes can be observed at the plaque edge.

Fibrous astrocytes are numerous, and a dense gliofibrillary network is distributed throughout the plaque replacing the myelinated tissue. The astrocytic fibers are often arranged in parallel rows creating an isomorphic gliotic scar (Fig. 3). Rosenthal fibers or corpora amylacea, degenerative products of a prolonged astroglial reaction, are often observed. Although the loss of myelin sheath with preservation of axons is the main neuropathological feature of demyelinating disorders, in chronic plaques axons are often disrupted and fragmented, and a few are in

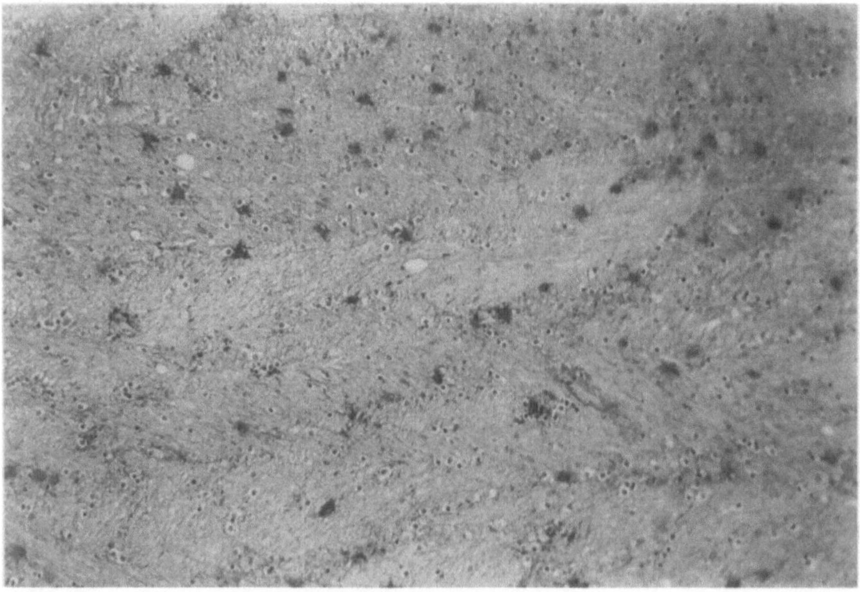

Fig. 3. Area of isomorphic gliosis in a deep white matter plaque

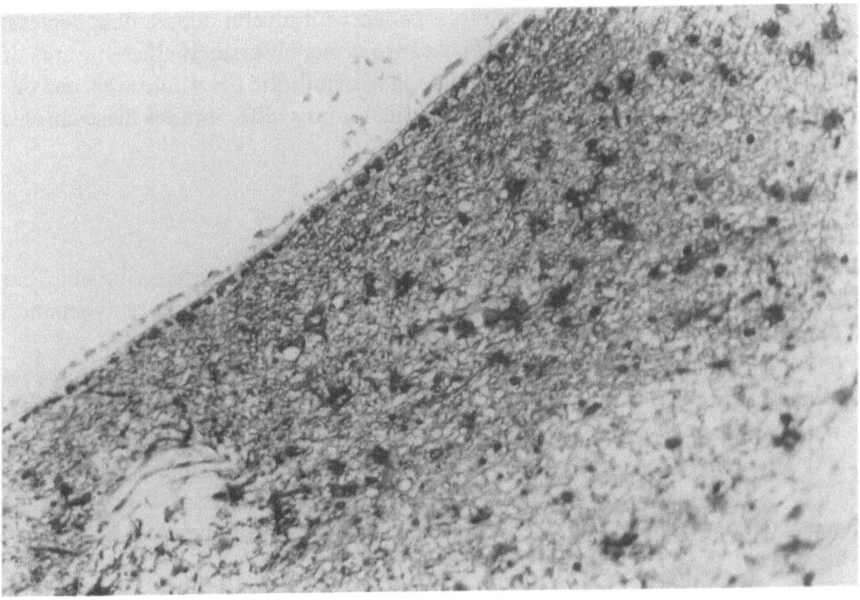

Fig. 4. Gliofibrillary network in the subependymal region

Wallerian degeneration. The loss of axons in the chronic plaques is probably the neuropathological counterpart of the neurological disability; possibly, the slowly evolving, chronic and progressive course of the disease is related to a continuous dropout of axons associated with fibrous astroglial replacement. Residual mild inflammatory activity can also be detected in chronic plaques. The demyelination process has stopped, but a few macrophages loaded with Oil Red O positive myelin debris can be observed in some enlarged Virchow Robin spaces or at the border of the lesion. The venules in the inactive plaque have hyalinized walls, and in the periventricular area fibrillar gliosis is usually intense (Fig. 4).

The ependymal cells can be injured and split with focal loss of ependyma replaced by a glial nodule (granular ependymitis). The ependymal granulation can give rise to important channels of entry in the CSF of IgG and other immune agents [6].

The chronic active plaque

In a chronic inactive plaque, inflammation and demyelination can reappear and these plaques have, therefore, the pathological features of both the old chronic and the acute/recent plaques. Perivascular inflammation, macrophages hyperplasia with surviving olygodendrocytes are evident at the edge of the plaque, which is usually markedly hypercellular and extending into the surrounding tissue. Therefore, the neuropathological characteristic of these plaques is a relatively inactive demyelinated and gliotic center, with an hypercellular border where the process of demyelination is active.

Astrocytic hyperplasia, macrophages loaded with myelin debris, lymphocytes, plasma cells and surviving oligodendrocytes are in fact observed in the same area. If the plaque is small, lymphocytes and macrophages infiltrate the whole area, and only the presence of an intense and diffuse fibrillar gliosis differentiates these chronic active plaques from the acute/recent lesions.

The shadow plaque

In the past years, repair of the demyelinated areas was not considered possible, but more recently partial remyelination of small plaques, or attempts at remyelination at the border of a chronic plaque, have been documented [7].

The cells responsible for the new myelin formation are not surviving oligodendrocytes; are probably oligodendrocytes which derive from the division and differentiation of precursor cells. However, the remyelination process is far from complete and the new myelin is usually thin, with short internodes. In coronal sections treated with stains for myelin the remyelinated plaques appear as areas of pale myelin staining (shadow plaques), which stand out from the normal surrounding white matter.

Immunohistochemistry

By immunohistochemistry a more functional *in situ* profile can be obtained [8]. In acute MS lesions CD4 T cells are the predominant population (Fig. 5), with CD8 T

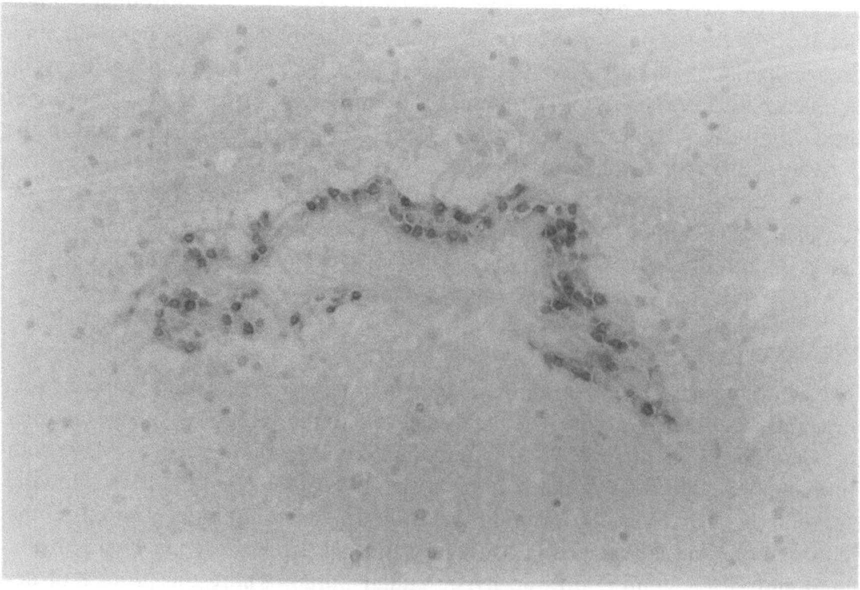

Fig. 5. Small perivascular cuff of CD4 positive T lymphocytes

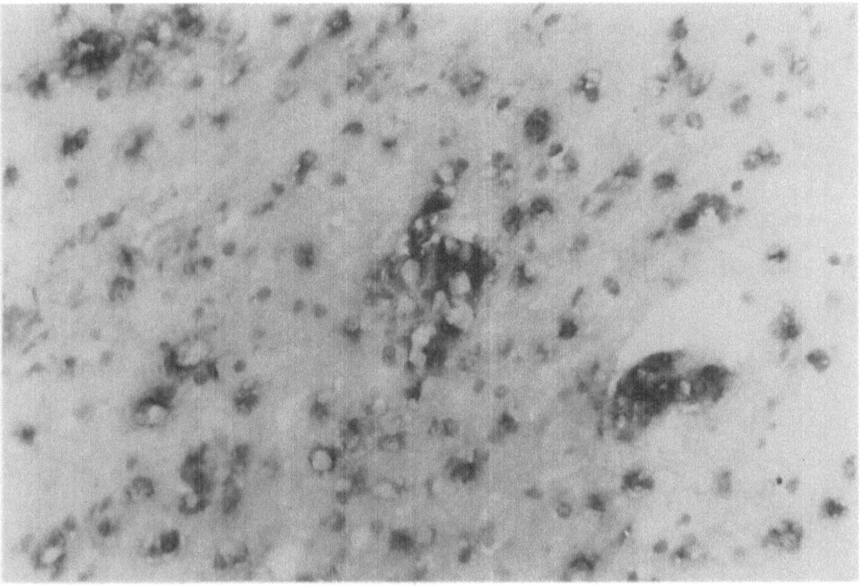

Fig. 6. Lesion edge of an acute/recent plaque showing macrophages Class II MHC positive

cells becoming more frequent in longstanding lesions. The majority of infiltrating lymphocytes in acute MS lesions bear TCR alpha beta; foamy macrophages stain for Class II MHC (Fig. 6).

In acute lesions more controversial is the presence of Class II MHC positive endothelial cells, while oligodendrocytes and reactive astrocytes are consistently negative for Class II staining. Microglia positive for HLA DR are seen at the plaque edge and in the surrounding apparently normal white matter. B cells are not a prominent cell population in the acute lesions, but plasma cells are seen particularly in the perivascular spaces.

In chronic active MS lesions, CD4 and CD8 can be detected in the lesion, in the perivascular areas (Fig. 7) and also in the surrounding apparently normal white matter. CD8 are more numerous at the edge of the lesion; macrophages and microglial cells are Class II MHC positive. It is unclear if endothelial cells are HLA DR positive because of the density of the infiltrates and the close contact between HLA DR positive macrophages and endothelial cells.

In chronic active and chronic silent lesions the majority of lymphocytes bear TCR alpha beta chain, but lymphocytes with TCR gamma delta can also be detected close to oligodendrocytes stained positively for a 65 KD heat shock protein (hsp 65), leading to the suggestion that chronic depletion of oligodendrocytes may be partially mediated by interaction between TCR of gamma delta T cells and hsp 65 [9].

A population of plasma cells is usually present not only in the perivascular space (Fig. 8), but also in areas of active myelin breakdown. IgG deposits have been

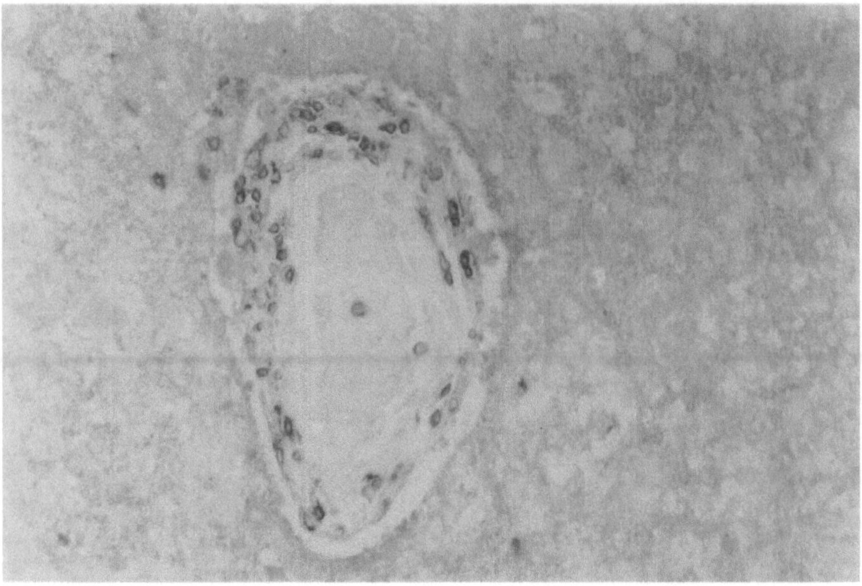

Fig. 7. A few lymphocytes infiltrate the vascular wall in a chronic/active MS plaque

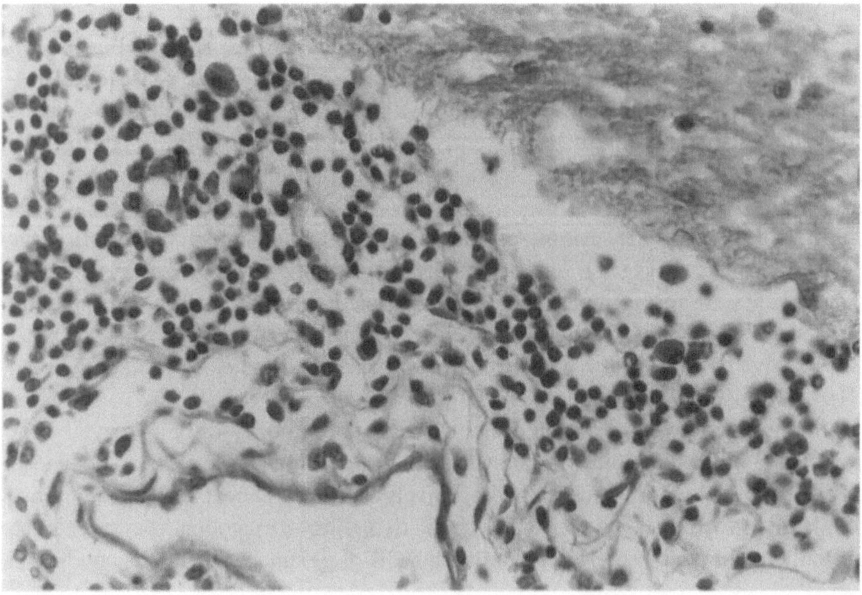

Fig. 8. Perivascular inflammation containing lymphocytes and plasma cells in a chronic plaque

demonstrated on macrophages, around blood vessels and in subependymal regions, and with higher density at the edge of the lesion. MBP specific antibody forming B cells have been also detected in chronic active areas; they are indicative of a substained B cell activation within the CNS of MS patients [10].

In the chronic silent plaques, scattered CD4 positive cells can be found throughout the lesion as well as within the normal CNS parenchyma, while CD8 T cells are absent from the lesion center. A moderate reactivity for IgG can be observed in the demyelinated area, and especially at the edge of the lesion.

Blood brain barrier, lymphocytes and cytokines

Blood brain barrier damage is a common finding in both acute and chronic lesions. Alpha 2 macroglobulin, Ig fractions, fibrinogen and complement fractions are commonly used as immunohistochemical markers of BBB breakdown. The anatomical basis of the BBB is provided by the "tight junctions" among cerebral endothelial cells.

In MS, autoaggressive T cells stick and adhere to the endothelial cells, and migrate in the Virchow Robin space, the interstitial space surrounding the small vessels in the parenchyma. It is an extension of the subpial, rather the subarachnoid space, lined by the basement membrane of the glia limitans peripherally, the outer surface of the blood lying centrally. This space is filled with fluid separated from CSF by the pia of the subarachnoid space; here, activated lymphocytes contact the adjacent microglia (pericytes). Pericytes, unlike perivascular cells, are also in close contact with glial cells, forming part of the glia limitans, which bounds internally the subarachnoid space where CSF is collected.

There is some evidence that expression of MHC antigens by pericytes may be constitutive and antigen presentation in the CNS could occur at the level of these cells. Antigen specific T cells are thought to recruit non-specifically activated T cells into the CNS via cytokine production and adhesion molecule expression. The immunocompetence of the interstitial Virchow Robin space can explain the possible detection of CSF pleiocytosis during the active phase of the disease, and the increased permeability of the BBB will allow antigen specific and antigen non-specific T cells to reach the CSF.

The majority of lymphocytes recruited in the CSF during an exacerbation are indeed activated, showing increased DNA synthesis [11] and IL-2 receptors [12]. All these data indicate that the increased cellular content of the CSF detectable in MS cases is derived from the inflammatory cells in the perivascular space.

Induction and activation of T lymphocytes require two signals from antigen presenting cells. The first signal is the binding of TCR to antigen-MHC complex and provides specificity [13]; the second is a non antigen specific "costimulatory" signal where the B7 costimulatory molecules appear critical. B7 immunoreactivity was detected on activated microglia and infiltrating macrophages within active lesions [14]; B71 and B72 were found upregulated with RT-PCR and immunocytochemistry in acute MS plaques [15].

The presence of costimulatory molecules in MS lesions could be critical for the induction of T cell responsiveness and the appearance of the autoimmune disease. B71 and B72, moreover, contribute to the activation of two different CD4 T cells, Th1 and Th2 [16]. These differing T cell populations produce different types of cytokines, namely Th1 IL2, TGF beta, INF gamma and Th2 IL4, IL5 and IL10. Using anti B71 antibodies the incidence of EAE is decreased while anti-B72 increases its severity [13]. It is possible that the maturation of CD4 helper precursors along the two alternative pathways to Th1 or Th2 phenotype is mediated by the costimulatory molecules B71 or B72. In MS it seems probable that the main T effector cells are of the Th1 phenotype.

Activation of T cells, after the recognition of the Ag bound to the MHC molecule and in the context of costimulatory molecules, results in proliferation and production of cytokines, the lymphocyte effector molecules. Elevated titers of cytokines in the CSF have been reported : interleukin 2, soluble IL2 and TNF alpha [17, 18], T cell activation antigen CD27 [19] . Olsson et al. [20] reported an increased frequency of CSF T cells secreting INF gamma in response to autoantigens such as like MBP or PLP, and elevated levels of TNF alpha in both CSF and serum. Significant correlations were found between CSF TNF alpha and clinical course.

Immunohistochemical studies demonstrated that cytokines were produced in the affected areas of the brain, at the edge of the lesions, and also in the proximal white matter. INF gamma is primarily produced by CD4 T lymphocytes of the T helper 1 phenotype. Both TNF alpha and the related cytokines or lymphotoxin (LT), which has been reported to kill oligodendrocytes by apoptosis, were identified in acute and chronic active lesions and were absent in chronic silent lesions [21]. LT was associated with T lymphocytes and microglia at the edge of the lesion, whereas TNF alpha was associated with astrocytes and foamy macrophages.

Recently, IL12 was found to be upregulated in acute MS plaques [15], and this increase was reported to be significant compared to the other cytokines. The main biological function of IL12 is to induce the differentiation of CD4 T helper into Th1, and the presence of IL12 in acute MS plaques could be relevant for the initiation of the Th1 response.

Immunoglobulins

The characteristic abnormality in the CSF of MS is the selective increase of IgG due to local synthesis within the CNS; CSF restricted oligoclonal IgG bands are present in up to 95% of MS cases. The specificity of these antibodies is still debated : some are directed against myelin and brain antigens such as glycolipids, PLP, MAG and MBP, and others against viral proteins, especially measles, rubella and herpes viruses. It is commonly thought that these antibodies are a nonsense response to antigens, irrelevant to the pathogenesis of MS, and that they are the product of plasma cell clones localized in the demyelinated plaques following the initial entry of antigen specific T cells.

However, humoral mechanisms are probably involved in myelin injury, and it is possible that the process of demyelination is due to the synergic action of macrophages, antibodies and complement. Lassmann et al. [22] were able to obtain large demyelinated lesions only in rats receiving simultaneously MBP, CD4 positive T cells and a monoclonal antibody against MOG antigen, a glycoprotein present on the myelin sheath and on the plasmalemma of oligodendrocytes. Moreover, Walsh and Tourtellotte [23], by studying with two dimensional electrophoresis CSF immunoglobulins and immunoglobulins eluted from MS brains, found that the oligoclonal bands in CSF were very stable over a long period of many years and that the Ig pattern, isolated from different regions of MS brains, and matched CSF were similar; this suggests that the humoral response in MS was uniform in each individual case and persists throughout the course of the disease.

These data contrast with the nonsense proposal of the origin and nature of *in situ* synthesized Ig in MS [23]. The sources of CSF Ig are the plasma cells, which are numerous in the chronic plaques and widely distributed also in the meninges and perivascular spaces [24]. Cells displaying the phenotype of circulating B cells are uncommon in acute and chronic lesions, and they probably differentiate into plasma cells in CNS parenchyma. A permanent population of plasma cells is therefore present in chronic lesions and their amount is considered sufficient for the detectable Ig in CSF.

Conclusions

A possible "scenario" of events in the developing of MS plaques starts with the crossing of the BBB by activated antigen specific Th1 lymphocytes. Inside the CNS parenchyma they produce INF gamma inducing MHC class II expression on microglial cells and pericytes. After the initial entry of T cells and the induction of ICAM-1 on endothelial cells, a second wave of non antigen specific activated T cells, along with macrophages, reaches the CNS.

This entry is accompanied by breakdown of the glia limitans, with an increased permeability of the BBB. Lymphocytes, activated lymphocytes and cytokines can be detected at CSF examination. At this stage, damage of myelin and oligodendrocytes occurs by release of toxic agents such as TNF alpha, LT and macrophages, and microglial cells remove the myelin sheaths.

Plasma cells migrate in the CNS and produce immunoglobulins which probably contribute to demyelination process. The inflammatory response then starts to be down regulated, but new inflammatory bursts occur. The CNS becomes increasingly populated by clones of plasma cells of restricted heterogeneity, which produce IgG, detected in the CSF and demonstrated as oligoclonal bands.

References

1. Schelling F (1986) Damaging venous reflux into the skull or spine: relevance to multiple sclerosis. Medical Hypotheses 21: 141-148

2. Peress NS, Siegelman J, Fleit HB (1987) High avidity periventricular IgG-Fc receptor activity in human and rabbit brain. Clin Immunol Immunopathol 42:229-238
3. Greenfield JG, Norman RM (1963) Demyelinating diseases. In: Blackwood W, McMenemey WH, Meyer A, Norman RM, Russell DS (eds) Greenfield's Neuropathology. 2nd ed, Edward Arnold, London, pp 475-519
4. Sanders V, Conrad AJ, Tourtellotte WW (1993) On classification of postmortem multiple sclerosis plaques for neuroscientists. J Neuroimmunol 46: 207-216
5. Raine CS, Traugott U (1990) The pathology of the myelinated axon. In: Adachi M and Sher JH (eds) Current Trends in Neurosciences-Neuromuscular diseases. Igaku-Shoin, New York, pp 229-275
6. Adams CWM, Abdulla YH, Torres EM, Poston RN (1987) Periventricular plaques in multiple sclerosis: their perivenous origin and relationship to granular ependymitis. Neuropathol Appl Neurobiol 13: 141-152
7. Prineas JW, Connell F (1979) Remyelination in multiple sclerosis. Ann Neurol 5: 22-31
8. Raine CS, Scheinberg LC (1988) On the immunopathology of plaque development and repair in multiple sclerosis. J Neuroimmunol 20: 189-201
9. Selmaj K, Brosnan CF, Raine CS (1991) Colocalization of lymphocytes bearing gamma delta receptor and heat shock protein hsp65 + oligodendrocytes in multiple sclerosis. Proc Natl Acad Sci 88: 6452-6456
10. Gerritse K, Deen C, Fasbender M, Ravid R, Boersma W, Claassen E(1994) The involvement of specific anti myelin basic protein antibody forming cells in multiple sclerosis immunopathology. J Neuroimmunol 49: 153-159
11. Noronha ABC, Richman DP, Arnason BGW (1980) Detection of in vivo stimulated cerebrospinal fluid lymphocytes by flow cytometry in patients with multiple sclerosis. New Eng J Med 303: 713-717
12. Tournier-Lasserve E, Lyon-Caen O, Roullet E, Bach MA (1987) IL 2 receptor and HLA Class 2 antigens on cerebrospinal fluid cells of patients with multiple sclerosis and other neurological diseases. Clin Exp Immunol 67: 581-586
13. Kuchroo VK, Das MP, Brown JA, Ranger AM, Zamvil SS, Sobel RA, Weiner HL, Nabavi N, Glimcher LH (1995) B71 and B72 costimulatory molecules activate differentially the Th1/Th2 developmental pathways: application to autoimmune disease therapy. Cell 80:707-718
14. De Simone R, Giampaolo A, Giometto B, Gallo P, Levi G, Peschle C, Aloisi F (1995) The costimulatory molecule B7 is expressed on human microglia in culture and in multiple sclerosis acute lesions. J Neuropathol Exp Neurol 54: 175-187
15. Hafler DA, Weiner HL (1995) Immunologic mechanisms and therapy in multiple sclerosis. Immunological Reviews 144: 75-107
16. Romagnani S (1994) Lymphokine production by human T cells in disease states. Ann Rev Immunol 12: 227-257
17. Gallo P, Piccinno M, Pagni S, Tavolato B (1988) Interleukin 2 levels in serum and cerebrospinal fluid of multiple sclerosis patients. Ann Neurol 24: 795-797
18. Trotter JL, Collins KG, Van der Veen R (1991) Serum cytokine levels in chronic progressive multiple sclerosis. Interleukin 2 levels parallel tumor necrosis factor alpha levels. J Neuroimmunol 33: 29-36
19. Hintzen RQ, Van Lier RAW, Kuijpers KC, Baars PA, Schaasberg W, Lucas CJ, Polman CH (1991) Elevated levels of a soluble form of the T cell activation antigen CD27 in cerebrospinal fluid of multiple sclerosis patients. J Neuroimmunol 35: 211- 217
20. Olsson T, Wang WZ, Hojeberg B, Kostulas V, Jiang YP, Andersson G, Ekre HR, Link KH (1990) Autoreactive T lymphocytes in multiple sclerosis determined by antigen-induced secretion of interferon gamma. J Clin Invest 86: 981-985
21. Selmaj K, Raine CS, Cannella B, Brosnan CF (1991) Identification of lymphotoxin and

tumor necrosis factor in multiple sclerosis lesions. J Clin Invest 87: 949-954
22. Lassmann H, Brunner C, Bradl M, Linington C (1988) Experimental allergic encephalomyelitis: the balance between encephalitogenic T lymphocytes and demyelinating antibodies determines size and structure of demyelinated lesions. Acta Neuropathol 75: 566-576
23. Walsh MJ, Tourtellotte WW (1986) Temporal invariance and clonal uniformity of brain and cerebrospinal IgG, IgA, and IgM in multiple sclerosis. J Exp Med 163: 41-53
24. Prineas JW, Wright RG (1978) Macrophages, lymphocytes and plasmacells in the perivascular compartment in chronic multiple sclerosis. Lab Invest 38: 409-421

The detection and significance of cerebrospinal fluid oligoclonal IgG

G. Giovannoni and E.J. Thompson

Introduction

Multiple sclerosis (MS) is characterised pathologically by inflammation, demyelination and variable degrees of axonal loss and gliosis. The inflammation tends to be perivascular and consists predominantly of macrophages, activated microglia, T and B lymphocytes, and occasional plasma cells. These plasma cells are responsible for the production of oligoclonal immunoglobulin (Ig) within the central nervous system (CNS), and the detection of this abnormal IgG in the cerebrospinal fluid (CSF) is useful as one of the laboratory criteria supporting the clinical diagnosis of MS.

Despite the discovery of cerebrospinal fluid (CSF) oligoclonal immunoglobulins in 1960 [1], their significance and role in the pathogenesis of this intriguing disease remains speculative. The diagnosis of MS remains clinical, but recent developments with new immunomodulatory therapies, such as interferon beta, are forcing a redefinition of the disease, with an increasing need for an earlier or even preclinical diagnosis. CSF analysis will play an important role in this regard, particularly the detection of oligoclonal IgG.

This chapter will review the methods of detection, the interpretation of results, the differential diagnosis, and the possible pathological significance of oligoclonal immunoglobulins in MS.

General

Basic science

The humoral response to a particular antigenic challenge is a complex process involving antigen recognition and the initial production of IgM. This IgM-producing step may be T cell independent, but it is followed by a process of immunoglobulin isotype switching and affinity maturation, which does require T cell help.

This T cell help involves extensive cross-talk between T and B cells via cell surface adhesion molecules and the production of specific cytokines. Briefly, the pro-

---

Department of Neuroimmunology, Institute of Neurology, Queen Square, London WC1N 3BG, UK

duction of immunoglobulins can be divided into the stages of B cell activation, proliferation and differentiation.

The activation of B cells occurs via antigen specific surface IgM receptors in the presence of the B cell activating cytokines IL1 and IL4. The antigen bound to the IgM receptors is then internalised, processed and presented in the context of MHC class II molecules. Proliferation, then, requires antigen-specific T cell help in the form of cytokines IL2, IL4 and IL5; this step requires the antigen-specific interactions between T and B cells via the TCR-MHC II receptor complex and important non-antigen specific co-stimulatory signals via other receptor ligand pairs like LFA1-ICAM1, LFA3-CD2 and CD28-B7. This results in T cell activation, and the production of the T cell cytokines that induce B cell proliferation and differentiation with isotype switching and affinity maturation. At this stage the type of T cell help influences the immunoglobulin isotype and subclass switching that occurs.

In general the Th2-like cytokine IL4 results in the production of IgE, IgG2 and IgG4, whereas TH1-like cytokines IL2 and INFγ induce IgG1 and IgG3 production. The net result of these processes is the formation of mature clones of plasma cells, each producing a specific antibody. This antibody is characterised by the combination of a single class of light and heavy chain having unique variable and hypervariable regions defining the molecules antigen specificity.

Antibody specificity and affinity

Antibody specificity can either be viewed as a measure of the goodness of fit between the antibody combining site (paratope) and the corresponding antigenic determinant (epitope), or the ability of the antibody to discriminate between similar or even dissimilar antigens. This binding specificity is associated with particular functional consequences, which are due to the properties conferred on the immunoglobulin molecule by its non-antigen binding sites, for example the ability to activate complement. The biological functions are isotype and subclass dependent.

In comparison to specificity, affinity of an antibody is a measure of the strength of the binding between antibody and antigen, such that a low affinity antibody binds weakly and high affinity antibody binds firmly. The process of affinity maturation preferentially selects for survival of plasma cells producing high affinity antibodies over those producing low affinity antibodies. Therefore, detecting high affinity antibodies to a specific antigen is a good indicator that the antigen is directly involved in driving the humoral response. Conversely, the presence of low affinity antibodies usually represents cross reactivity or an anamnestic response.

Type of humoral response

Using electrophoresis it is also possible to classify the humoral response according to the number of antibody clones produced: a monoclonal antibody results from a single plasma cell clone, oligoclonal due to several clones, or polyclonal representing a general increase in immunoglobulin production with no specific discernible clones noted above the background.

Monoclonal - A monoclonal response can represent the initial stage of an oligoclonal response as the other antibody clones are not yet visible, but more commonly it represents a single abnormal clone of plasma cells associated with one of the plasma cell dyscrasias. Difficulty arises when using methods which are very sensitive, such as iso-electric focusing (IEF), since they may detect bands invisible by methods with a lower resolution, such as agarose electrophoresis.

The natural history of an IEF detected gammopathy, which is not detectable on agarose electrophoresis, a relatively common dilemma, is unknown. Until more is known about the natural history of these low level gammopathies, we recommend that routine serum IEF and electrophoresis with immunofixation is performed at 6 to 12 monthly intervals, applying the same principles used for the monitoring of a monoclonal gammopathy of undetermined significance (MGUS).

Oligoclonal - The oligoclonal response, however, is a well recognised phenomenon and represents an immunological response to a specific antigen or set of antigens, and is associated with numerous infectious, autoimmune and inflammatory conditions (Table 1). This response, when it is generated against a set of known antigens, in viral encephalitis for example, can be used as a specific diagnostic tool. When the eliciting antigens are not known, as in MS, the presence of an oligoclonal pattern is non-specific, representing a local humoral response, and is not helpful as a definitive diagnostic aid.

Polyclonal - This is a non-specific increase in immunoglobulin synthesis, and is commonly associated with systemic diseases with an immunological response to numer-

Table 1. Inflammatory diseases of the CNS associated with CSF oligoclonal IgG bands

Disorder	Approximate incidence of oligoclonal bands (%)
Multiple sclerosis	95
Auto-immune	
• Neuro-SLE	50
• Neuro-Behcet's	20
• Neuro-sarcoid	40
• Harada's meningitis-uveitis	60
Infectious	
• Acute viral encephalitis (< 7 days)	< 5
• Acute bacterial meningitis (< 7 days)	< 5
• Subacute sclerosing panencephalitis (SSPE)	100
• Progressive rubella panencephalitis	100
• Neurosyphilis	95
• Neuro-AIDS	80
• Neuro-borelliosis	80
Tumour	< 5
Hereditary	
• Ataxia-telangectasia	60
• Adrenoleukodystrophy	100

ous antigens, for example a portal-systemic shunt due to underlying liver disease. This is not a pattern commonly associated with neurological diseases.

Methods of detection

Historically, an increase in CSF immunoglobulin was detected by the colloidal gold Lange curve. This was replaced by agarose and polyacrylamide gel electrophoresis, and later by nitro-cellulose transfer and immunofixation of these electrophoresis techniques. All these methods have now been superseded by the development of the more sensitive technique of IEF mentioned above.

Briefly, this technique uses a pH gradient to separate IgG populations on the basis of charge; these are then transferred onto a nitro-cellulose membrane and immunostained using an anti-human immunoglobulin (Fig. 1). Problems with interpretation can arise when inhomogeneities in the ampholytes used in establishing the pH gradients cause artefactual bands: these must be recognised. It must be stressed that this is a qualitative test and subjected to observer bias, and should therefore always be read by an observer blinded to any other clinical information.

IEF patterns

As CSF is an ultrafiltrate of plasma, it contains immunoglobulins which are passively transferred from the plasma, as well as any immunoglobulins which maybe synthesised locally. Any systemic pattern of immunoglobulin production seen in plasma or serum will therefore be mirrored in the CSF. It is imperative that any CSF analysis for oligoclonal bands is accompanied by a paired blood analysis. Fig. 1 represents the various combinations of patterns found in the CSF and serum, together with our current interpretation (Table 2).

Differential diagnosis

Any neurological disease characterised by an intrathecal inflammatory response can produce an oligoclonal IgG response. Table 1 provides a list, with an approximate proportion of cases having oligoclonal bands.

Multiple Sclerosis

Diagnosis

Quantitative versus qualitative assessment - Although all the main immunoglobulin isotypes are synthesised intrathecally in MS, IgM and IgG remain the most useful diagnostically. The presence of CSF oligoclonal IgG has a higher diagnostic sensitivity than any of the quantitative indices demonstrating abnormal blood CSF barrier function or increased intrathecal immunoglobulin synthesis in MS (Table 3). It is because of this that a European consensus report has recommended the use of IEF with im-

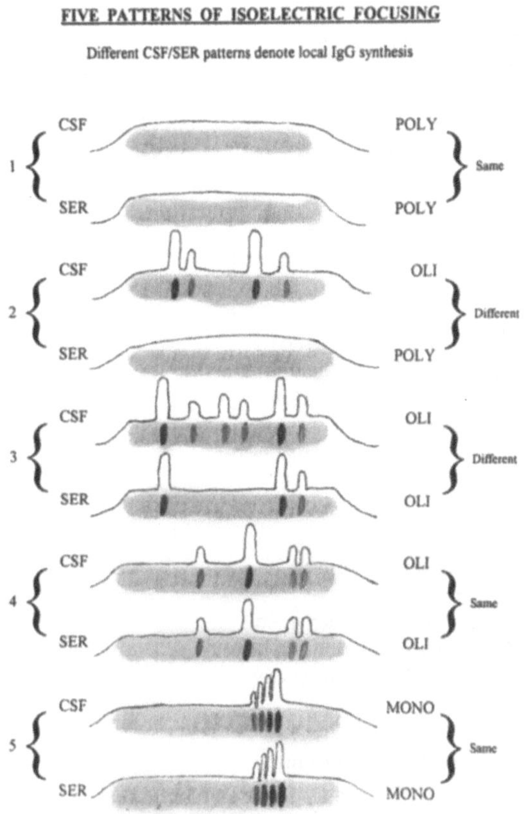

a

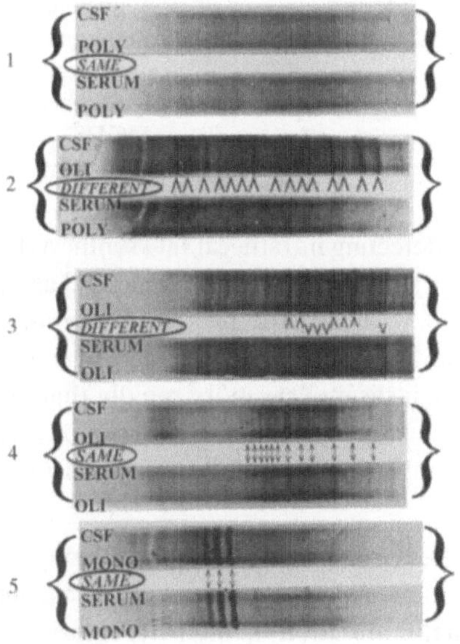

b

Figs. 1a, b. A schematic diagram (**a**), and the corresponding immunoblots (**b**) of the various CSF and serum isoelectric focusing patterns. The interpretation of these patterns is presented in Table 2 (Reproduced with permission of the publishers of the Journal of Neurology, Neurosurgery and Psychiatry)

POLY = Polyclonal
OLI = Oligoclonal
MONO = Monoclonal
Cathode is on the right

Table 2. Interpretation of IEF patterns (see Fig. 1)

Pattern	**Interpretation**	**Examples**
Type 1	Normal	
Type 2	Oligoclonal IgG is present in the CSF with no apparent corresponding abnormality in serum, indicating local intrathecal synthesis of IgG	Multiple sclerosis
Type 3 (or greater than pattern)	There are IgG bands in both the CSF and serum, with additional bands present in the CSF. The oligoclonal bands which are common to both CSF and serum imply a systemic inflammatory response, whilst the bands which are restricted to the CNS suggest that there is an additional CNS-only response	Multiple sclerosis, SLE, sarcoid, etc.
Type 4 (or mirror pattern)	There are oligoclonal bands present in the CSF, which are identical to those in serum. This is not indicative of local synthesis, but rather the pattern is consistent with passive transfer of oligoclonal IgG from a systemic inflammatory response	Guillain-Barré syndrome
Type 5	There is a monoclonal IgG pattern in both CSF and serum, the source of which lies outside the CNS	Myeloma, monoclonal gammopathy of undetermined significance (MGUS)

munodetection as the method of choice for detecting intrathecal IgG synthesis [2].

Oligoclonal IgM is possibly slightly more sensitive than oligoclonal IgG as a predictor for the development of MS, but it is more difficult to detect, and currently the methods for its detection have not been standardised. Oligoclonal IgM is however useful in infectious diseases of the CNS, as it disappears more rapidly than oligoclonal IgG on the removal of the antigenic challenge, and is useful in demonstrating a response to therapy. In contrast, oligoclonal IgG persists for many years and possibly for life after disease-specific therapy.

Light chains - During the synthesis of the heavy and light chain components of immunoglobulins, κ and λ light chains are produced in excess. As these light chains also undergo gene rearrangement and somatic mutation of their hypervariable regions, they produce a response that can also be classified as mono, oligo, or poly-

Table 3. Quantitative vs. qualitative CSF abnormalities in MS

Test	% Abnormal
Quantitative	
Abnormal blood CSF barrier function (Albumin quotient > 7 x 10^{-3})	12%
Increased IgG quotient (IgG index > 0.88)	70-80%
Increased cell count (> 4/ µl)	50%
Qualitative	
IEF - oligoclonal bands	95%

clonal. This free light chain response can be measured using polyacrylamide gel electrophoresis (PAGE), and has similar diagnostic utility as the oligoclonal IgG and M responses. The quantity of free light chain produced has also been used as an index of plasma cell activity. Patients with more active MS have greater quantities of free light chains in their CSF than patients with less active disease. Excess urinary free light chains have also been described in patients with MS, indicating excessive plasma cell activity.

Sensitivity, specificity and positive predictive value - Oligoclonal IgG is a very sensitive test in MS, occurring in approximately 95% of cases with clinically definite MS [3]. It lacks specificity however, as it occurs in numerous other diseases, some of which can masquerade as MS. The positive predictive value of oligoclonal bands in the diagnosis of MS will vary from laboratory to laboratory, depending on the incidence of MS in the population as well as the referral pattern of CSF specimens. In a specialised CSF laboratory this positive predictive value was 66% when all CSF specimens received were analysed [3].

Early diagnosis of MS - As new disease-modifying therapies for MS become available, it is becoming more important to diagnosis MS earlier, and possibly even in the preclinical stage. The CSF oligoclonal IgG and M response is very useful in achieving this aim, and is able to predict with reasonable certainty those who will develop clinically definitive MS. This is important as these patients may benefit from early institution of disease modifying therapies. For example, patients who develop isolated syndromes associated with the later development of MS, who have either oligoclonal IgG or M bands in their CSF, have a 78-89% chance of developing MS within 18 months [4]. Similarly, an asymptomatic identical twin whose sibling has MS has approximately 25-30% chance of developing clinically definite MS. If this twin has CSF oligoclonal bands and MRI changes suggestive of MS, the chances will be much higher. Subjects like this may become suitable candidates for disease modifying therapy.

CSF oligoclonal IgG versus MRI and evoked potentials - CSF oligoclonal bands are found in approximately 90-95% of patients with clinically definite MS, which compares favourably to MRI. Evoked potentials, however, have a slightly lower yield with approximately 80% of patients having at least one abnormal result in either the visual, auditory brainstem, or somatosensory pathways. It must be stressed that each of these paraclinical tests assesses a different aspect of the disease process, and they should therefore be viewed as complementary investigations providing important but different diagnostic information.

Change in oligoclonal banding pattern with disease progression

Some patients with MS, found to be oligoclonal negative at their initial presentation, become oligoclonal positive later on in their disease course [5]. Similarly, patients with MS of relatively recent onset are more likely to develop increased numbers of CSF bands as their disease progresses. This is in contrast to the stable pattern found in patients with disease of long duration [6].

These interesting observations suggest that the number of antigenic epitopes inducing the oligoclonal response increases with time. Further evidence supporting this is the increased frequency of specific antibodies to myelin oligodendrocyte glycoprotein in patients in the progressive phase of their illness. Some Authors hypothesise that this represents a specific shift in the immunological target towards myelin specific antigens and hence the oligodendrocyte, and may account for the transition from relapsing to a progressive course [7]. The oligoclonal response not only has temporal variability, but has also been elegantly shown to vary spatially, with the elutes from individual MS plaques of the same patient demonstrating different banding patterns [8]. The implications of these findings are enormous and suggest that many antigens are driving the immunological response in MS at a focal, rather than at a systemic level; this also supports the concept that antigenic spreading may occur in MS. This is a phenomenon described in animal models of autoimmune disease in which the disease inciting antigenic epitope generates numerous other autoimmune responses to either new epitopes within the inciting antigen (intramolecular antigenic spread), or to other distinct organ-specific antigens (intermolecular antigenic spread) [9].

Antibody specificity and affinity in MS

In MS the oligoclonal IgG response has been found to react to numerous common viral antigens. By using a thiocynate titration, Luxton et al. showed that these antibodies are of low affinity and probably unrelated to the primary antigenic stimulus. This is in contrast to viral encephalitis in which the oligoclonal antibody responses to the specific antigens are of higher affinity. Interestingly, in encephalitis antibody responses to other antigens unrelated to the offending organism occur, but these are low affinity, and represent cross-reactivity or an anamnestic response. This anamnestic or secondary response presumably occurs from the non-specific activation of B cells in the favourable immunological milieu provided by an inflamed CNS.

Despite ongoing and extensive efforts in MS, no specific antigen has yet been found against which a high affinity primary oligoclonal response occurs.

IgG subclasses and complement fixation

Although all subclasses of IgG are represented in the oligoclonal response, they are predominantly IgG1 and IgG3, and these have complement fixing properties. The ability of these antibodies to fix complement may play an important role in oligodendrocyte toxicity and demyelination, as oligodendrocytes are very susceptible to the membrane attack complex of complement.

Complement also aids in myelin opsonisation, augmenting its phagocytosis which appears to be mediated via immunoglobulin Fc and complement receptors. Increased levels of activated components of complement, C3a and C5a, are found in the CSF of patients with MS.

Anti-myelin antibodies

Models of experimental allergic encephalomyelitis (EAE) have demonstrated the importance of specific humoral anti-myelin responses in the induction of demyelination. In certain models of EAE the coadministration of anti-myelin oligodendrocyte glycoprotein (MOG) antibodies induces striking demyelination with a pathological pattern not too dissimilar from MS [10]. *In vitro*, anti-myelin antibodies greatly augment the process of myelin phagocytosis, via Fc receptor-mediated uptake.

Electron microscopy of pathological specimens from patients with MS show that myelin phagocytosis occurs via clatharin-coated pits, a process associated with Fc receptor recognition [11]. It therefore appears that the myelin phagocytosis by microglia and macrophages is an immunoglobulin-assisted process, and that a specific anti-myelin humoral response may be involved in the pathogenesis of demyelination.

Serum oligoclonal bands

Although the majority of patients with MS have an IEF pattern demonstrating intrathecal synthesis only, approximately 40% of patients show some serum oligoclonal bands [12]. These are usually fewer and fainter than the CSF bands, and they are both concordant and disconcordant, representing identical as well as different clones to those found in the CNS compartment respectively. This implies an additional and separate systemic immunological response as intrathecally-produced immunoglobulins passing back into the blood would be of insufficient quantity to be detected in the serum by IEF. This raises interesting questions about the nature of this systemic response, and whether these patients have a different disease or a more generalised dysregulation of the immune system. The presence of a systemic response in MS is associated with elevated levels of intrathecal synthesis of IgG, increasing age, later disease onset, and the presence of serum auto-antibodies.

Oligoclonal negative MS

The diagnosis of MS must be seriously reconsidered in the absence of a CSF oligoclonal immunoglobulin response, as less than 5% of patients with clinically definite MS are oligoclonal negative [5]. These patients may have MS of recent onset who have yet to fully express the local oligoclonal immunoglobulin response, or they may represent a specific subtype of MS characterised by a more benign course with less disability and lower MRI lesion loads [5]. Some oligoclonal negative patients given the clinical diagnosis of MS prove to have other pathological diagnoses.

Disease monitoring

The oligoclonal IgG response currently has no role in the monitoring of disease activity in MS. On the other hand, the quantitative and qualitative assessment of CSF free light chains, an index of excessive plasma cell activity, has been associated with disease activity [13]. Unfortunately, repeat CSF free light chain measurements are unlikely to gain widespread support, as there is a general resistance from both physicians and patients to performing repeated lumbar punctures.

Conclusions

The detection of a CSF oligoclonal IgG response by IEF is a non-specific but sensitive aid in the diagnosis of MS, and should be used after the exclusion of other infective and inflammatory disorders. It is due to local or intrathecal IgG production and provides information about the immunopathogenesis of MS. It should be used in parallel with evoked potentials and MRI in aiding the clinical diagnosis of MS, as each set of investigations provides different information about the pathogenesis of MS.

Importantly, the CSF oligoclonal IgG response is not only a diagnostic but a predictive test, allowing one to assign to a patient with an isolated syndrome, a risk of developing MS. This will almost certainly aid in the selection of patients for earlier disease-modifying therapy.

References

1. Lowenthal A, van Sande M, Karcher D (1960) The differential diagnosis of neurological diseases by fractionating electrophoretically the CSF γ-globulins. J Neurochem 6:51-5
2. Andersson M, Alvarez-Cermeño J, Bernadi G et al (1994) Cerebrospinal fluid in the diagnosis of multiple sclerosis: a consensus report. J Neurol Neurosurg Psychiatry 57:897-902
3. McLean BN, Luxton RW, Thompson EJ (1990) A study of immunoglobulin G in the cerebrospinal fluid of 1007 patients with suspected neurological disease using isoelectric focusing and the log IgG-index. Brain 113:1269-1289
4. Sharief MK, Thompson EJ (1991) The predictive value of intrathecal immunoglobulin

synthesis and magnetic resonance imaging in acute isolated syndromes for the subsequent development of multiple sclerosis. Ann Neurol 29:147-151
5. Zeman AZJ, Kidd D, McLean BN et al (1996) A study of oligoclonal band negative multiple sclerosis. J Neurol Neurosurg Psychiatry 60: 27-30
6. Thompson EJ, Kaufmann P, Rudge P (1983) Sequential changes in oligoclonal patterns during the course of multiple sclerosis. J Neurol Neurosurg Psychiatry 46:115-118
7. Lassmann H, Suchanek G, Ozawa K (1994) Histopathology and the blood-cerebrospinal fluid barrier in multiple sclerosis. Ann Neurol 36:S42-46
8. Mattson DH, Roos RP, Arnason GW (1980) Isoelectric focusing of IgG eluted from multiple sclerosis and subacute sclerosing panencephalitis brains. Nature 287:335-337
9. Lehmann PV, Sercarz EE, Forsthuber T, Dayan CM, Gammon G. (1993) Determinant spreading and the dynamics of the autoimmune T-cell repertoire. Immunol Today 14:203-207
10. Wekerle H, Kojima K, Lannes-Vieira J, Lassman H, Linington C (1994) Animal models. Ann Neurol 36:S47-53
11. Raine CS, Scheinberg LC (1988) On the immunopathology of plaque development and repair in multiple sclerosis. J Neuroimmunol 20:189-201
12. Zeman AZJ, Keir G, Luxton R, Thompson EJ (1996) Serum oligoclonal bands are a common and persistent feature of multiple sclerosis and have a systemic origin. Q J Med 89: 187-193
13. Vakaet A, Thompson EJ (1985) Free light chains in the cerebrospinal fluid: an indicator of recent immunological stimulation. J Neurol Neurosurg Psychiatry 48:995-998

The detection of IgA, IgM and free light chains abnormalities in cerebrospinal fluid

C.J.M. Sindic

Introduction

The discrete bands observed in the cathodic region of agar gel electrophoresis in most CSF from MS patients [1] represent mainly IgG produced by a limited number of B-cell clones. The term *oligoclonal* as proposed by Laterre in 1965 [2] is now widely accepted to define these bands present in CSF but not in blood. Already in 1977, some of these bands have been identified as free kappa or lambda light chains [3]. In addition, we know that a minority of these IgG bands are actually directed against various neurotropic viruses, such as measles, rubella, varicella zoster and mumps [4]. All these infections are very common during childhood and probably affect the brain more frequently than clinically observed. They probably induce the recruitment of specific B-cells within the central nervous system (CNS) and their persistence as memory cells. The MS process could lead to the reactivation of these memory B-cells and to a "polyspecific" production of antibodies against some of these neurotropic viruses.

However, more than 30 years after the discovery of these CSF restricted oligoclonal IgG bands, we still do not understand their physiopathological significance, if any. All attempts at defining MS specific antigens responsible for this humoral immune response have been unsuccessful. This lead several Authors to study other immunoglobulin (Ig) isotypes (IgA and IgM), and the frequent occurrence of free light chains in the CSF from MS patients, in order to shed some light on the role of this humoral immune response in MS process.

The IgA in MS CSF

IgA is present in normal human serum at about one-fifth of the concentration of IgG, but is catabolized around five times faster than IgG, indicating a similar rate of synthesis. IgA is also the most abundant Ig in secretions. Serum IgA is predominantly monomeric (about 87%), and is mainly produced in bone marrow, while in external secretions most of the locally produced IgA is polymeric. IgA is present in

---

Laboratory of Neurochemistry, Catholic University of Louvain, Avenue Mounier 5359, B-1200 Brussels, Belgium

normal CSF at about one-tenth of the concentration of IgG, and is mostly (> 95%) monomeric, as the particular function of the blood-CSF and the blood-brain barriers leads to the relative exclusion from CSF of proteins of high molecular weight [5].

When IgA is intrathecally synthesized as, for example, during the course of herpetic encephalitis or tuberculous meningitis, we have previously shown [5] that the proportion of polymeric IgA increased dramatically in the CSF. This may be explained by the fact that peripheral blood lymphocytes (PBL) secrete about equal proportions of monomeric and polymeric IgA, either spontaneously or after mitogenic stimulation. Since the lymphocytes invading the subarachnoid space during an intrathecal immune reaction have a blood origin, one may expect that CSF IgA, when locally produced, will have the characteristics of the IgA secreted by PBL, rather than those of serum IgA.

This fact should be kept in mind in any trial to quantify the intrathecal IgA synthesis. For example, the calculation of an IgA quotient or an IgA index, based on total IgA levels assayed in serum and CSF, takes into account neither the relative proportions of monomeric and polymeric IgA in both fluids, nor the underestimation of polymeric CSF IgA by most immunoassays (except for immunonephelometry). The current quantitative formulae are rather a crude approach in the assessment of IgA local synthesis.

Nevertheless, to simplify matters we calculated an IgA index based on the total IgA levels [6]. In a reference group of 50 non-neurological patients, this index displayed a logarithmic normal distribution, with a mean value of 0.23 and an upper reference limit (mean + 2 SD) of 0.42. Only 16 out of 117 MS patients (14%) showed an IgA index higher than this value of 0.42 [6]. This percentage is much lower than that of 41% reported by Lolli et al. [7]. The discrepancy may be due to the use of a lower reference value of the IgA index (mean: 0.15; upper limit: 0.34) by these Authors.

The occurrence of CSF-restricted oligoclonal IgA bands was reported in 16 out of 20 MS patients by Grimaldi et al. [8], but in only one out of 39 by Link and Laurenzi [9], and this has been disputed [10]. By means of an immunoaffinity-mediated capillary blot technique of higher sensitivity, we failed in a recent study [11] to detect a significant proportion of CSF samples containing oligoclonal IgA bands. Only 5 samples out of 124, (including 2 out of 33 MS patients) contained such bands, in spite of the fact that an intrathecal IgA production was thought to occur in 19 samples (15% of the total group; 6% in the MS group) on the basis of Reiber's graph [12]. The local production of IgA seems to be therefore mostly polyclonal. This is also exemplified by the study of 5 cases of herpetic encephalitis: in all cases we observed the intrathecal synthesis of specific anti-herpes simplex virus (HSV) IgA antibodies, which displayed infrequently an oligoclonal banding always superimposed on a predominant polyclonal background.

As the locally produced IgA probably originated from a restricted number of B-cell clones present within the CNS and was actually oligoclonal, the diffuse *polyclonal* pattern observed on immunoblots could be due to post-translational modification. Different glycosylations could specifically be responsible for a microheterogeneity that we do not detect by current techniques.

In conclusion, IgA analysis, either by quantitative or qualitative techniques, is of little value for a laboratory-supported diagnosis of MS, and does not appear to be relevant to the immune pathogenesis of the MS process. Very rare cases of MS, however, may be characterized by a significant intrathecal IgA production, although always associated with the presence of CSF-specific oligoclonal IgG (A. Sena, personal communication).

The IgM in MS CSF

As IgM is normally the first antibody to be produced in immune reactions, the assessment of its intrathecal synthesis may be central to the early diagnosis of inflammatory and infectious disorders of the CNS. Because of its large molecular size, IgM diffuses from blood into the CSF at a very slow rate. Its concentration in the normal CSF is about 10,000 times lower than in blood [13]. Any increase of the CSF levels will reflect either an increase in the transudation through the blood-CSF barrier, or a local production. The latter could be assessed either by quantitative methods (for example, the IgM index [13], the Reiber's formula and graph [12] etc.), or by qualitative methods demonstrating an electrophoretic pattern of CSF IgM, that differs from the one observed in the corresponding serum [14].

Very sensitive methods such as radioimmunoassays, enzyme-immunoassays, or particle-counting immunoassays are required to detect the trace amounts of IgM present in the normal CSF. In a group of 50 non-neurological patients, the mean CSF IgM level, calculated on log values, was 130 μg/1, with an upper reference limit of 380 μg/1 (mean + 2 SD). The IgM index also had a logarithmic normal distribution, with an upper reference limit of 0.079 [13]. In a cohort of 160 MS patients, 42 (26%) had an elevated IgM index (Table 1) [6]; this index was more frequently increased in patients in relapses (28 out of 94 or 30%) than in patients in remission (2 out of 15 or 13%), but this difference was not statistically significant [15].

In a more recent study of 46 other MS patients [16], the IgM index was increased in 18 cases (39%), and also more frequently in acute relapses: 11 out of 21 (52%) vs 7 out of 25 (28%) with other forms of the disease. Among the 13 patients experiencing the first bout of their disease, 7 (54%) displayed an increased IgM index. One must keep in mind that the well documented subclinical activity of MS as demonstrated by MRI makes partly illusory a reliable discrimination of relapsing from remitting disease on clinical grounds.

The validity of the IgM index in the assessment of the intrathecal synthesis of this isotype has frequently been criticized. Indeed, the IgM index has been shown to increase with progressive dysfunction of the blood-CSF barrier [17], and in such cases the index may yield falsely abnormal results. In MS, however, the blood-CSF barrier is either normal or minimally disturbed, and the concordance between an increased IgM index and the presence of oligoclonal IgM bands is high (see below).

We have recently used an immunoaffinity-mediated capillary blot technique for the detection of CSF-restricted oligoclonal IgM bands in a group of 46 MS patients [16]. The overall sensitivity of the immunoblot assay was increased by the glu-

Table 1. IgA, IgM and free light chains abnormalities in MS CSF (personal results)

High IgA index [6, 11]	14% (16 of 117 patients) 6% (2 of 33 patients)
Oligoclonal IgA bands [11]	6% (2 of 33 patients)
High IgM index [13, 16]	26% (42 of 160 patients) 39% (18 of 46 patients) (52% during acute relapses)
Oligoclonal IgM bands [16]	28% (13 of 46 patients) 43% (during acute relapses)
High free kappa levels High free lambda levels [26]	81% (155 of 191 patients) 92% (176 of 191 patients)
High free kappa index High free lambda index [26]	86% (123 of 143 patients) 86% (86 of 101 patients)
Oligoclonal free kappa bands Oligoglonal free lambda bands [27]	92% (44 of 48 patients) 67% (33 of 48 patients)

taraldehyde fixation of the primary antibody-antigen complex. The pH gradient was extended between 4 and 6 by a mixture of ampholines, and samples were pretreated by dithiotreitol in order to obtain a better IgM penetration and migration into the gel (Fig. 1). Finally, the specificity was increased by the double use of anti-IgM (μ-chain specific) antisera, first on the polyvinylidene difluoride (PVDF) sheet for the immunoaffinity transfer, and then for immunodetection with the peroxidase complex.

In CSF samples with a normal IgM concentration (<0.38 μg/ml) no IgM was detectable on the immunoblot because the limit of detection was found to be 6 ng in 15 μl of samples put on the gel, corresponding to an IgM concentration of about 0.40 μg/ml. This was the case for 21 out of the 46 MS patients (Table 2). In 12 patients, the staining of the immunoblot revealed a diffuse polyclonal background identical in CSF and serum. No intrathecal production was demonstrated in such cases, and an increased transudation through the blood-CSF barrier was assumed to be present, as all these samples showed high IgM concentrations.

Finally, 13 MS patients (28%) displayed CSF-restricted oligoclonal IgM bands, most often superimposed on a polyclonal background. These IgM bands were detectable in 9 out of 21 patients with acute relapses, but in only 4 out of the 25 patients with the other forms of the disease. Among the 13 patients experiencing the first bout of their disease, 6 displayed CSF oligoclonal IgM bands (Table 2).

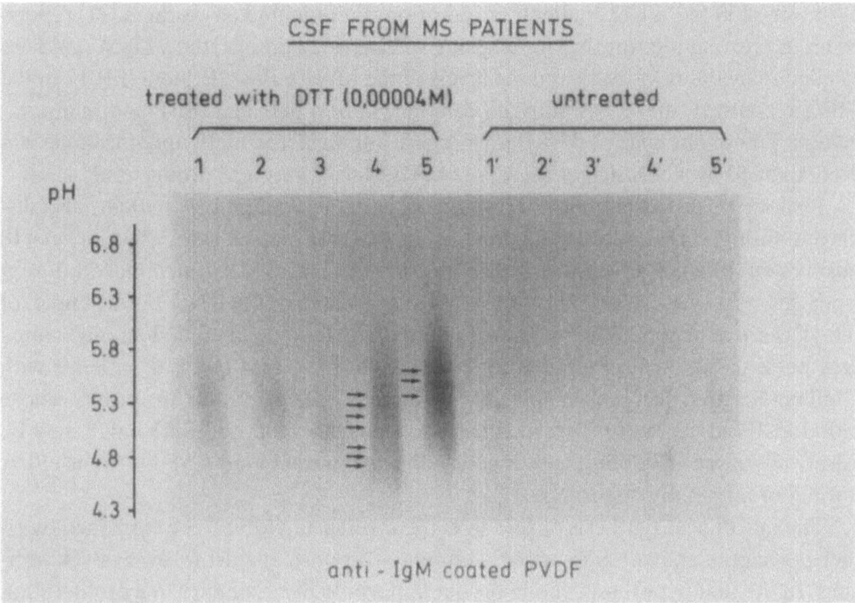

Fig. 1. Immunoblots of IgM in the CSF from 5 MS patients, either treated or untreated with dithiotreitol (DTT: 0.04 mM). The DTT reduction is necessary to obtain a migration of the IgM into the gel before its transfer on the PVDF sheet. Note the presence of oligoclonal IgM bands in the CSF of 2 MS patients (lanes 4 and 5) superimposed on a polyclonal background detectable in the 5 samples

These results are in accordance with those reported by other Authors using various methods [18- 21]. As Sharief and Thompson [19], we find that the detection of oligoclonal IgM bands varies with the clinical activity of the disease. Specifically, oligoclonal IgM bands were present in about 45% of first bouts of the disease as well as in acute relapses. In initial MS, Lolli et al. [20] also found an increased IgM index

Table 2. Patterns of IgM immunoblots in MS

Clinical course[a]	Not detectable	Polyclonal	Oligoclonal
Stable (N = 7)	1	5	1
Acute relapses (N = 21)[b]	8	4	9
Secondary progressive (N = 6)	5	–	1
Primary progressive (N = 12)	7	3	2
Total (N = 46)	21	12	13

[a] In absence of any treatment, except for 4 patients with a secondary progressive course who received previously cyclophosphamide
[b] Including the first bout of the disease in 13 cases

in 14 out of 33 (42%) MS patients. In addition, Frequin and co-workers [21] reported an interesting relationship in relapsing MS between an intrathecal IgM synthesis calculated by the IgM index and the levels of the Myelin Basic Protein (MBP) in the CSF. In contrast, there was no significant correlation between MBP and IgG parameters. Treatment by high dose intravenously administered methylprednisolone was accompanied by a correlative decrease of MBP levels and IgM production.

It shoud be noted that all 13 MS patients with oligoclonal IgM patients also displayed an high IgM index, the frequency of the latter being somewhat higher (18 positive out of 46, that is 39%). This high concordance in MS in strikingly different from that observed in patients with infectious diseases of the CNS. For example, of the 15 cases of aseptic meningitis with a high IgM index, only 3 had an oligoclonal IgM pattern. Similarly, we failed to detect oligoclonal IgM bands in patients with Guillain-Barré syndrome, in spite of the increase of IgM index in 4 cases whose blood-CSF barrier was severely impaired. In such a case, the IgM index may be falsely increased [17], but its use is acceptable when the blood-CSF barrier is either normal or minimally disturbed.

The intrathecal synthesis of IgM in early or relapsing MS can be explained by either polyclonal activation or specific immune response against foreign or self antigens. In the first hypothesis, the Ig production would be secondary to a proliferation of B lymphocytes under the mitogenic action of unknown endogenous or exogenous factors, and would not play any pathogenic role in MS.

In such a case, one would expect to find auto-antibodies, such as IgM rheumatoid factors. However, we failed to detect the latter in the CSF of 80 MS patients, using a sensitive instrumental agglutination test [13]. As IgM antibodies may also specifically react against polysaccharidic moieties of glycoproteins, we also looked for possible auto-antigenic target(s) within the minor glycoprotein fraction of the myelin proteins. These were purified by lentil-lectin chromatography from homogenates of human brain myelin, and used on antigen-specific immunoblots, with negative results to date (unpublished data).

Free light chains in MS CSF

In vitro stimulated B-cells and plasma cells synthesize and release free light chains. Excess of free light chains has been detected in sera from patients with diseases such as chronic lymphocytic leukaemia, multiple myeloma and Sjögren's syndrome, which are all accompanied by increased synthesis of immunoglobulins. The presence of free light chains in CSF from MS patients has been reported as a *double ring* formation in radial immunodiffusion already in 1974 [22], and oligoclonal bands of free light chains were described 3 years later [3]. Only few studies dealt with quantitative analysis of the free light chains in the CSF [23-26].

By means of a particle counting immunoassay [26], we detected free lambda chains even in normal CSF (mean ± SD: 120 + 60 ng/ml), whereas free kappa chains were undetectable in most controls (detection limit: 25 ng/ml). Increased levels of free kappa and lambda chains were observed in the CSF from 81% and 92% MS patients, respectively, but also in a significant proportion (>50%) of patients with other neurological, especially infectious, disorders (Table 1).

Our results disagreed with those of Rudick and co-workers [23-25], who proposed that hight CSF levels of free kappa and lambda chains were specific to MS and acute CNS infections, respectively. However, in these studies, the serum levels of the free light chains were not taken into account for the assessment of a transudation process. Indeed, the use of a free light chain index increased dramatically the specificity of these assays: the free kappa chain index was increased (>5.5) in 86% of MS patients, and in 40% of patients with CNS infections; a free lambda index higher than 17 (upper reference level) was also observed in 86% of MS patients, but in only 23% of patients with infectious diseases (Table 1).

In MS, high levels of free kappa and lambda indices correlated significantly ($p < 0.01$) with either the presence of oligoclonal IgG bands or a high IgG index.

The possible advantages of the detection of oligoclonal free light chains versus oligoclonal IgG bands may be summarized as follows:

1) because of a lower molecular weight, free light chains may diffuse more easily into the subarachnoid space;
2) in contrast, IgG may be absorbed by interacting with membranes (myelin) via their Fc fragments and with antigens via their F (ab') 2 fragments;
3) from a technical point of view, immunoblots of free light chains are easy to interpret because of the absence of a polyclonal background.

By means of an immunoaffinity-mediated capillary blot technique, we obtained a very high sensitivity, with a limit of detection of 20 ng of free kappa or lambda Bence-Jones proteins, and a very high specificity without interference by light chains bound to immunoglobulins [27].

In our hands, up to 92% of MS patients display CSF-restricted oligoclonal free kappa bands, and 64% oligoclonal free lambda bands. In most cases, these bands are numerous (up to 12), very sharp and not superimposed on a polyclonal background (Fig. 2). They are also detectable on a broader pH range than IgG bands, but are never observed in the corresponding sera, nor in the CSF and sera from patients without neurological disorders. In some MS cases, free light chains bands may be observed in CSF samples devoid of detectable oligoclonal IgG bands, as also reported by others [28].

It should also be noted that in this study [27] all CSF samples with increased absolute levels of free kappa or free lambda chains displayed oligoclonal bands of the corresponding free light chains. In addition, free light chain bands were sometimes detectable without an increase in their absolute levels. These bands represent, therefore, a sensitive marker of an intrathecal immune response especially in MS. However, their significance remains obscure.

At variance with another report [29], we failed to find significant differences in the kappa and lambda indices between patients with active disease and patients in apparent clinical remission.

It is unlikely that the presence of these free light chains is related to the degradation of Ig, as free heavy chains (gamma, alpha, mu) were never found in the CSF. These light chains could be involved in the idiotypic regulation of the humoral re-

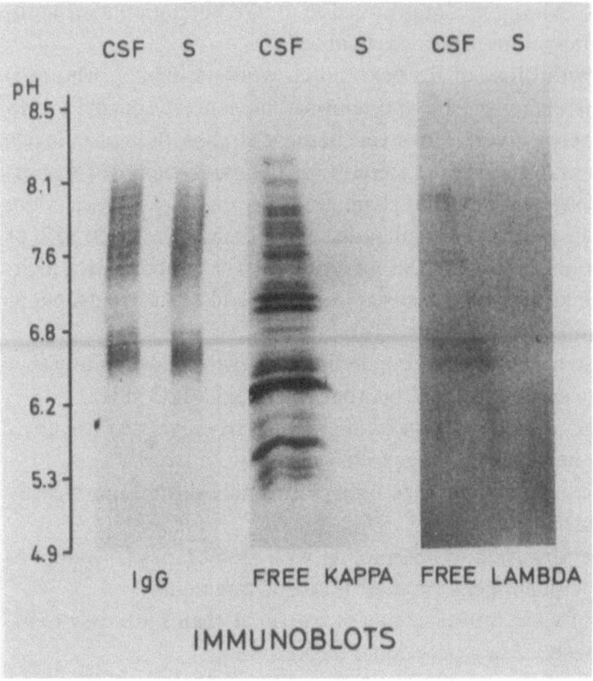

Fig. 2. Immunoblots of IgG, free kappa and free lambda chains from unconcentrated MS CSF and the corresponding serum diluted to the same IgG concentration, after simultaneous isoelectric focusing in agarose gel. Note the presence of several oligoclonal IgG and free lambda bands, and of numerous oligoclonal free kappa bands

sponse, as shown in a murine model [30]. From a practical point of view, their detection as oligoclonal bands is a complementary test to establish a laboratory-supported diagnosis of MS. A recent paper by K. Lamers et al. confirms and extends our previous results [32].

Conclusions

Most of MS patients display a long-standing humoral immune reaction within the CNS, which leads mainly to the local synthesis of oligoclonal IgG and free light chains (Table 1).

A significant proportion of patients (25% to 40%) also has an intrathecal IgM response. The IgA production is less frequently observed (6% to 14%) but, as already discussed, its frequency is likely underestimated by quantitative assays and formulae. These immune parameters do not clearly correlate with clinically detectable disease activity. The only exception could be the local IgM response, which seems to be more pronounced in early or relapsing MS. It is still not known whether this local immune response is an epiphenomenon or plays a key role in the MS process. So far, no definitive evidence supports the assumption that the locally produced Ig antibodies are against MS-specific antigen(s). Are these immunoglobulins produced at random (*non sense antibodies*), or is it linked to the demyelinating process? In absence of any definite answer, two observations may be made:

a) in a very elegant model of experimental autoimmune encephalitis studied by Lassmann et al. [31] large demyelinated lesions occurred only in rats receiving simultaneously myelin basic protein-specific, CD4 positive T cells and a monoclonal antibody against a glycoprotein present on both the myelin sheath and the plasma membrane of the oligodendrocyte (the MOG antigen);

b) in early MS lesions, only T cells and macrophages are present, but no B-cells; oligodendrocytes survive and even proliferate during this early stage. In contrast, in late chronic MS, oligodendrocytes are destroyed, and B-cells are detectable in the perivascular spaces. The neuropathological features of the MS process are thus changing in the course of the disease, and these changes may be due to the additional presence of (auto)-antibodies.

References

1. Lowenthal A, Vansande M, Karcher D (1960) The differential diagnosis of neurological diseases by fractionating electrophoretically the CSF gamma-globulins. J Neurochem 6: 51-56
2. Laterre EC (1965) Les protéines du liquide céphalo-rachidien à l'état normal et pathologique. Thèse, Arscia-Maloine, Paris
3. Vandvik B (1977) Oligoclonal IgG and free light chains in the cerebrospinal fluid of patients with multiple sclerosis and infectious diseases of the central nervous system. Scand J Immunol 6:913-922
4. Sindic CJM, Monteyne Ph, Laterre EC (1994) The intrathecal synthesis of virus-specific oligoclonal IgG in multiple sclerosis. J Neuroimmunol 54:75-80
5. Sindic CJM, Delacroix DL, Vaerman JP, Laterre EC, Masson PL (1984) Study of IgA in the cerebrospinal fluid of neurological patients with special reference to size, subclass and local production. J Neuroimmunol 7:65-75
6. Sindic CJM, Boon L, Chalon MP, Laterre EC (1987) Intrathecal synthesis of immunoglobulins in MS patients. In: Lowenthal A, Raus J (eds) Cellular and humoral immunological components of cerebrospinal fluid in multiple sclerosis. Plenum Publishing Corporation, pp 47-59
7. Lolli F, Halawa I, Link H (1989) Intrathecal synthesis of IgG, IgA, IgM and IgD in untreated multiple sclerosis and controls. Acta Neurol Scand 80:238-247
8. Grimaldi LME, Roos RP, Nalefski EA, Arnason BGW (1985) Oligoclonal IgA bands in multiple sclerosis and subacute sclerosing panencephalitis. Neurology 35:813-817
9. Link H, Laurenzi MA (1979) Immunoglobulin class and light chains type of oligoclonal bands in CSF in multiple sclerosis determined by agarose gel electrophoresis and immunofixation. Ann Neurol 6:107-110
10. Mehta PD, Patrick BA, Miller JA (1983) Absence of oligoclonal IgA in CSF and serum of multiple sclerosis patients. J Neuroimmunol 6:67-69
11. Sindic CJM, Monteyne Ph, Bigaignon G, Laterre EC (1994) Polyclonal and oligoclonal IgA synthesis in the cerebrospinal fluid of neurological patients: an immunoaffinity-mediated capillary blot study. J Neuroimmunol 49:109-114
12. Reiber H, Felgenhauer K (1986) Protein transfer at the blood cerebrospinal fluid barrier and the quantitation of the humoral immune response within the central nervous system. Clin Chim Acta 163:319-328
13. Sindic CJM, Cambiasco CL, Depre A, Laterre EC, Masson PL (1982) The concentration of IgM in the cerebrospinal fluid of neurological patients. J Neurol Sci 55:339-350

14. Keir G, Walker RWH, Thompson EJ (1982) Oligoclonal immunoglobulin M in cerebrospinal fluid from multiple sclerosis patients. J Neurol Sci 57:281-285
15. Sindic CJM Cerebrospinal fluid proteins in diseases of the nervous system. Thesis - Catholic University of Louvain, Faculty of Medicine, Brussels
16. Sindic CJM, Monteyne Ph, Laterre EC (1994) Occurrence of oligoclonal IgM bands in the cerebrospinal fluid of neurological patients: an immunoaffinity-mediated capillary blot study. J Neurol Sci 124:215-219
17. Felgenhauer K (1982) Differentiation of the humoral immune response in inflammatory diseases of the central nervous system. J Neurol 228:223-237
18. Sharief MK, Keir G, Thompson EJ (1990) Intrathecal synthesis of IgM in neurological diseases: a comparison between detection of oligoclonal bands and quantitative estimation. J Neurol Sci 96:131-142
19. Sharief MK, Thompson EJ (1991) Intrathecal immunoglobulin M synthesis in multiple sclerosis. Relationship with clinical and cerebrospinal fluid parameters. Brain 114:181-195
20. Lolli F, Siracusa G, Amato MP, Fratiglioni L, Dal Pozzo G, Galli E, Amaducci L (1991) Intrathecal synthesis of free immunoglobulin light chains and IgM in initial multiple sclerosis. Acta Neurol Scand 83:239-243
21. Frequin STFM, Barkhof F, Lamers KJB, Hommes OR, Borm GF (1992) Cerebrospinal fluid myelin basic protein, IgG and IgM levels in 101 multiple sclerosis patients before and after treatment with high dose intravenous methylprednisolone. Acta Neurol Scand 86:291-297
22. Iwashita H, Grunwald F, Bauer H (1974) Double ring formation in single radial immunodiffusion for kappa chains in multiple sclerosis cerebrospinal fluid. J Neurol 207:45-52
23. Rudick RA, Peter DR, Bidlack JM, Knutson DW (1985) Multiple sclerosis: free light chains in cerebrospinal fluid. Neurology 35:1443-1449
24. Rudick RA, Pallant A, Bidlack JM, Herndon RM (1986) Free kappa light chains in multiple sclerosis spinal fluid. Ann Neurol 20:63-69
25. De Carli C, Menegus MA, Rudick RA (1987) Free light chains in multiple sclerosis and infections of the CNS. Neurology 37:1334-1338
26. Fagnart OC, Sindic CJM, Laterre EC (1988) Free kappa and lambda light chain levels in the cerebrospinal fluid of patients with multiple sclerosis and other neurological diseases. J Neuroimmunol 19:119-132
27. Sindic CJM, Laterre EC (1991) Oligoclonal free kappa and lambda bands in the cerebrospinal fluid of patients with multiple sclerosis and other neurological diseases. J Neuroimmunol 33:63-72
28. Gallo P, Tavolato B, Bergenbrant S, Siden A (1989) Immunoglobulin light chain patterns in the cerebrospinal fluid. A study with special reference to the occurrence of free light chains in cerebrospinal fluid with and without oligoclonal immunoglobulin G. J Neurol Sci 94:241-253
29. Vakaet A, Thompson EJ (1985) Free light chains in the cerebrospinal fluid: an indicator of recent immunological stimulation. J Neurol Neurosurg Psychiatry 48:995-998
30. Puccetti A, Koizumi T, Migliorini P, André-Schwartz J, Barrett KJ, Schwartz RS (1990) An immunoglobulin light chain from a lupus-prone mouse induces autoantibodies in normal mice. J Exp Med 171:1919-1930
31. Lassmann H, Brunner C, Bradl M, Linington C (1988) Experimental allergic encephalomyelitis: the balance between encephalitogenic T lymphocytes and demyelinating antibodies determines size and structure of demyelinated lesions. Acta Neuropathol 75:566-576
32. K. Lamers et al (1995) Cerebrospinal fluid free kappa light chains versus IgG findings in neurological disorders: qualitative and quantitative measurements. J Neuroimmunol 62:19-25

Evaluation of blood cerebrospinal fluid barrier function and quantification of the humoral immune response within the central nervous system

H. Reiber

The detection of disease-related changes in the brain is the basic target of cerebrospinal fluid (CSF) analysis. Besides cell count and cell differentiation, basic information is provided from proteins in CSF, in particular by detection of intrathecally synthesized immunoglobulin fractions in CSF. For the most sensitive quantification of a brain-derived CSF protein fraction, besides the corresponding blood-derived fraction in CSF we need to take into account the blood CSF barrier function and in particular the variations of CSF flow rate.

This chapter describes the physiology and biophysics of the blood CSF barrier function for proteins, the hyperbolic discrimination function in CSF/serum quotient diagrams, the neuroimmunological response patterns of IgG, IgA and IgM and their diagnosis-related relevance.

The functional definition of the blood CSF barrier

CSF flow and protein diffusion - Physiology

CSF is produced in the choroid plexus of the ventricles as a water-like fluid with few cells (0-4 cells/μL), low protein content (0.2% of blood total protein) and salt concentrations comparable to blood [1-3]. CSF flows through the cisternae, divides into either a cortical and a lumbar branch of subarachnoid space, and drains through arachnoid villi into venous blood without any filtration (Fig. 1). CSF flow and CSF production rates were considered forty years ago by Davson (cited in [1]).

The largest protein fractions in lumbar CSF originate from blood. The serum proteins enter CSF via passive diffusion continously along its way between ventricles and lumbar subarachnoid space, increasing the total protein content about 2.5 fold (Fig. 1). The time for serum proteins to equilibrate between blood and CSF (about one day for albumin and several days for IgM) as well as the blood: CSF concentration gradient (200:1 for albumin and 4000:1 for IgM) depend on the molecular size of the single protein. Blood-derived and brain-derived protein concentrations in CSF are modulated by normal and pathological variations of CSF flow rate

Laboratory of Neurochemistry, University of Göttingen, Robert-Koch-Str. 40, D - 37075 Göttingen, Germany

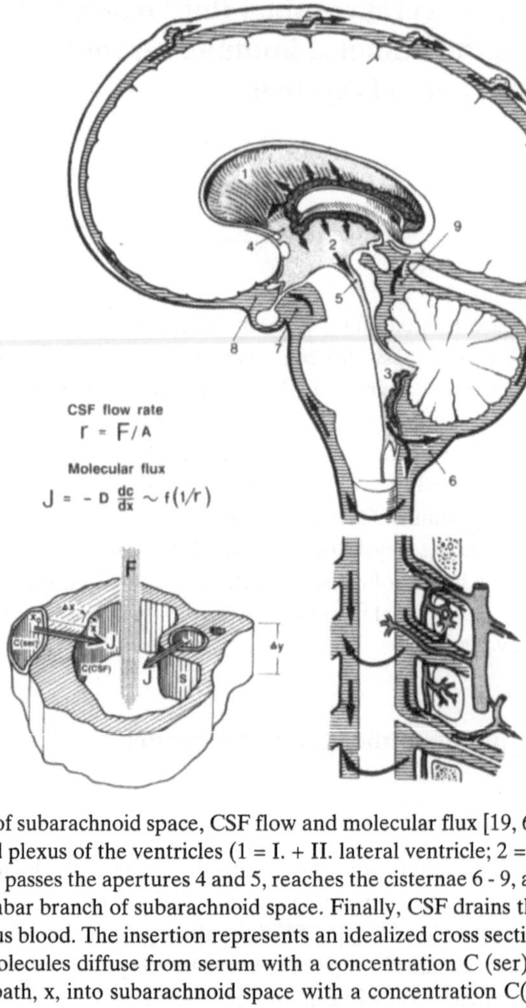

Fig. 1. Model of subarachnoid space, CSF flow and molecular flux [19, 6]. After CSF production in choroid plexus of the ventricles (1 = I. + II. lateral ventricle; 2 = III. ventricle; 3 = IV. ventricle) CSF passes the apertures 4 and 5, reaches the cisternae 6 - 9, and divides into a cortical and a lumbar branch of subarachnoid space. Finally, CSF drains through the arachnoid villi into venous blood. The insertion represents an idealized cross section through subarachnoid space. Molecules diffuse from serum with a concentration C (ser) through tissue along the diffusion path, x, into subarachnoid space with a concentration C(CSF). The molecular flux, J, depends on the local gradient $\Delta c/\Delta x$ or dc/dx and the diffusion constant, D. The CSF concentration increases with decreasing volume exchange, i.e. decreasing CSF volume bulk flow ($F \approx 500$ mL/day). The flow rate of a molecule in CSF is $r = F/A$ with A, the varying cross section of subarachnoid space

[4]. The diffusion-dependent molecular flux, J, the CSF volume flow, F, and their interaction (Fig. 1) are the main parameters which determine the concentration of each CSF protein, and thus define the blood CSF barrier functionally [4-6].

This definition represents the new paradigm of blood CSF barrier function. The many different morphological structures, passed by a protein on its way from blood at different sites into CSF, are treated in a unique diffusion model as a mean effective diffusion pathlength [4-6]. In contrast, for the treatment of the blood brain barrier [7] we regard the details of morphological structures.

The bood CSF barrier function for low molecular weight compounds

Aminoacids with a facilitated, transporter-mediated transfer [8], or glucose and Vitamin C [9] with active transport from blood into CSF have to be treated differently from a diffusion model for proteins.

Quantitation of the blood CSF barrier function by the albumin CSF/serum quotient

The albumin CSF/serum concentration quotient Q_{Alb} = Alb(CSF)/Alb(serum) represents the generally accepted quantitative measure for the blood CSF barrier function. Albumin, as a protein exclusively synthesized outside the brain, represents an ideal parameter to characterize all the influences and restrictions to a protein on its way between serum and lumbar CSF, including diffusion barriers and CSF flow rate. In general, by formation of a CSF/serum concentration quotient, the variation of the individual serum concentation is eliminated, i.e. we get a normalized CSF concentration as a dimensionless quotient with values between 0 and 1.

The reference range of the albumin quotient [10] is age dependent, with a most dramatic change in the first four months of life. Due to the rostro-caudal concentration gradient in CSF, the albumin quotient increases from ventricular via cisternal to lumbar CSF [11] (see Fig. 5). Increased albumin quotients correlate with the extent of blood CSF barrier dysfunction, i.e. with the reduction of CSF flow rate in a non-linear function [4]. Referring other CSF/serum protein concentration quotients (Q_{IgG}, Q_{IgA} Q_{IgM}) to the albumin quotient as the barrier function [4, 10] provides a most valuable way of discriminating a brain-derived from a blood-derived CSF protein fraction.

Pathological blood CSF barrier function means reduced CSF turnover

Many neurological diseases are concomitant with an increase in CSF protein concentrations together with a change of concentration ratios between single proteins (e.g. Q_{IgG}: Q_{Alb}). Earlier interpretations of this fact suggested a barrier "leakage" to explain an increase of permeability and a failure of selectivity, with a loss of molecular size-dependent discrimination for protein transfer between blood and CSF.

The careful analysis of CSF protein data from 4300 patients [4] contradicts the view of morphological changes as cause of increased CSF protein concentrations. The reduced CSF flow rate is sufficient to explain the observed change in CSF concentrations of single proteins in neurological diseases [4]. A decreased CSF turnover may be caused by a reduced production rate of CSF, by flow restriction in the subarachnoid space or by restricted passage through the arachnoid villi. The corresponding physiological aspects are described in a paragraph below.

As reviewed in the next paragraph, this new paradigm of the blood-CSF barrier dysfunction [4-6] could be derived from the laws of diffusion, with a complete explanation of changing CSF protein ratio without any morphological changes of the barrier-related structures: the molecular flux of proteins into CSF as molecular size-dependent diffusion makes for the "selectivity", the molecular size-dependent discrimination by the barrier function. The CSF flow rate also modulates the molecular flux, and thus changes the CSF concentration ratios of proteins with different size (e.g. Q_{IgG}: Q_{Alb}) in the sense of a hyperbolic function [4]. The empirical fit of the hyperbolic functions with the large figure of patient data {4] led to the actual CSF/serum quotient diagrams [10] for discrimination of brain-derived and blood-derived CSF protein fractions and for characterization of intrathecal immunological reaction patterns in neurological diseases.

Physiology and pathophysiology of CSF and blood CSF barrier function
1. Cerebrospinal fluid (CSF) originates from choroid plexus in the ventricles, flows through cisternae and subarachnoid space and finally drains through the arachnoid villi into venous blood. CSF flow starts around time of birth and reaches maximum rate about 4 months after birth, with complete maturation of the arachnoid villi. 2. Blood proteins enter into CSF along its way between ventricles and lumbar subarachnoid space, inducing a 2.5 fold increase of total protein concentration between ventricular and lumbar CSF. 3. Protein transfer from brain into CSF and from blood into CSF follows the laws of diffusion as a function of molecular size. The diffusion-related transfer of proteins into CSF is the cause for molecular size-dependent discrimination, i.e. selectivity, of the barrier function. As a consequence, we have larger CSF/serum quotiens for the smaller molecules $Q_{Alb} > Q_{IgG} > Q_{IgA} > Q_{IgM}$. Again, the smaller albumin molecule equilibrates faster between blood and CSF than the larger molecules IgG, IgA or IgM. 4. The absolute concentration of blood-derived CSF proteins is modulated by CSF flow rate. In general, the actual CSF protein concentrations are determined by: individual serum concentration and individual diffusion pathway, age, site of puncture and volume of extraction. 5. The blood CSF barrier for proteins is defined functionally by non-linear interaction of molecular flux and CSF flow rate. 6. A blood CSF barrier dysfunction means decreased CSF flow rate. The structures along the diffusion pathway for proteins between blood and CSF are conserved (no "leakage"). 7. A blood CSF barrier dysfunction, i.e. pathologically reduced CSF flow, can have different causes: reduced CSF production rate, restricted flow in subarachnoid space and restricted passage through arachnoid villi.

The molecular flux / CSF flow model of blood-CSF barrier function

Non-linear dynamics versus leakage model

The following arguments, summarized from [4], indicate that the observed dynamic of CSF protein concentrations and the change of their ratios under pathological conditions occur with conservation of barrier-related structures. There is no real loss of selectivity or change in permeability constants in blood CSF barrier dysfunction associated with neurological diseases.

In Fig. 2, with the most severe blood-CSF barrier dysfunctions where CSF protein concentrations approach 20% of blood concentrations, the mean IgG, IgA and IgM quotients do not approach each other. There remains molecular size dependent discrimination in spite of decreasing ratio QlgG: QAlb, etc. This statistical view can

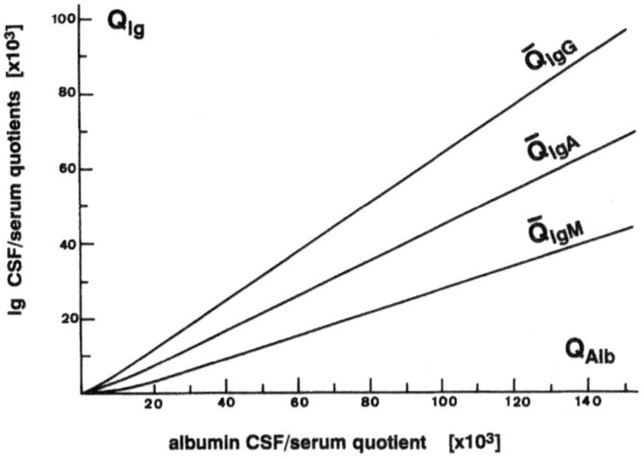

Fig. 2. Mean immunoglobulin quotients $\bar{Q}_{IgG}$, $\bar{Q}_{IgA}$, $\bar{Q}_{IgM}$ as a function of an increasing albumin CSF/serum concentration quotient Q_{Alb}. The mean values are calculated from groups of patients (populations) with similar albumin quotients of varying Q_{Alb} intervals (total n = 4300 patients). The mean values were fitted with a hyperbolic function [4, 5]

be supported by a kinetic approach shown in Table 1. The data from a very early first lumbar puncture and a later repeated puncture of the same patient with bacterial meningitis represent a type of natural stopped flow experiment. It is shown that there are larger relaxation times for the protein molecules of larger size to reach a new steady state concentration in CSF. At the first puncture, the albumin CSF concentration reached 47% of the value for day two, compared to 21%, 12% and only 5% for IgG, IgA and IgM values, respectively. The smaller molecule albumin appeared twice, 4 times or 10 times faster in lumbar CSF than the larger molecules IgG, IgA or IgM. This observation is contradictory to a "leakage" model of barrier dysfunction. There are other strong arguments for the conservation of blood CSF barrier "structures" relevant to protein diffusion from blood to CSF, shown in Fig. 3 and Table 2. With most severe blood CSF barrier dysfunction there is no change in the population variation coefficient, $\Delta Q/\bar{Q}$ (see Fig. 3). This means that each population of patients with a common extent of barrier dysfunction, e.g. at $Q_{Alb} = 10$, $Q_{Alb} = 50$ or $Q_{Alb} = 150 \cdot 10^{-3}$, has the same biological variation of "selectivity" around the mean. This is possible only if the structures on the pathway, relevant for diffusion, remain constant. In addition, the constant population variation coefficients of IgG, IgA and IgM in CSF do not approach the corresponding blood varia-

Table 1. Kinetics of CSF/serum concentration quotients after onset of disease

	QAlb ($\cdot 10^3$)	**QIgG** ($\cdot 10^3$)	**QIgA** ($\cdot 10^3$)	**QIgM** ($\cdot 10^3$)	**Cell count**
1st day	146	42	22	5	872/µL
2nd day	311	203	184	105	154000/µL

A patient with bacterial meningitis was punctured on the 1st and 2nd day after the onset of clinical symptoms. Both punctures were done before onset of a local IgG, IgA or IgM synthesis (no detectable oligoclonal IgG fractions from brain). The increasing cell count in CSF is not the source of a local immunoglobulin synthesis

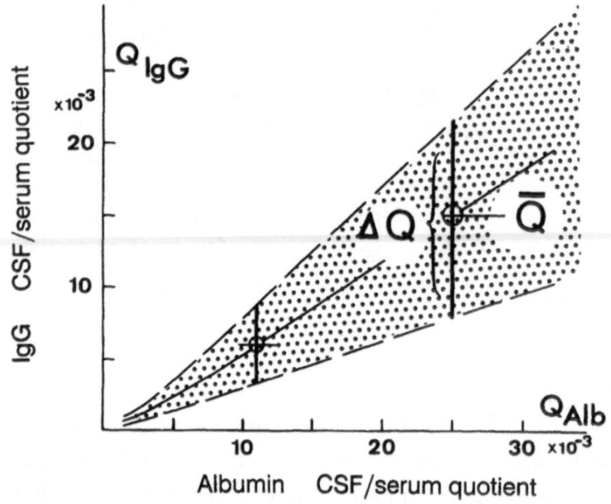

Fig. 3. Definition of the population variation coefficient ($\Delta Q/\bar{Q}$) for the immunoglobulin CSF/serum quotients. For example, patients with an albumin quotient of $Q_{Alb} = 25 \cdot 10^{-3}$ might have an IgG quotient Q_{IgG} between 8 and $21.6 \cdot 10^{-3}$. This range describes the difference between upper and lower border line involving 99% of the cases by analogy to the mean ± standard deviations in case of a Gaussian distribution. Independent of each other we find $\Delta Q_{IgG} = 13.6 \cdot 10^{-3}$ (≈ 6SD) and $\bar{Q}_{IgG} = 15 \cdot 10^{-3}$. The coefficient of variation for the particular population can be calculated as $\Delta Q_{IgG}/\bar{Q}_{IgG} = 13.6/15 = 0.91$

Table 2. Comparison of population variation coefficients for CSF/serum concentration quotients ($\Delta Q/\bar{Q}$) and serum concentrations (6 SD/$\bar{x}$)* in the same group of patients (n = 4300 patients)

		IgG	IgA	IgM
CSF	($\Delta Q/\bar{Q}$)	0.91	1.35	2.0
Blood	(6 SD/$\bar{x}$)*	1.8	3.3	3.5

* 6 SD/x means 6 standard deviations (SD) divided by the mean of the population (x) which corresponds to the definition of the population variation coefficient including 99% of the population

tion coefficients (Table 2), postulated to occur in a leakage model with (partial or complete) loss of selectivity.

The hyperbolic function: a theoretical implication of the molecular flux / CSF flow model

With the observed constancy of morphological conditions for protein diffusion between blood and CSF, it was possible to apply a diffusion model with the following boundary conditions:

1. the concentration of a single protein at one side of the barrier (e.g. in the blood) remains constant, i.e. the reservoir of proteins is infinite;
2. proteins diffuse through tissue, which does not need to be characterized with its very different structures like endothelial cell layer with tight junctions, fenestrat-

ed capillaries, intercellular fluid, ependymal cell layer with and without tight junctions. This simplification is possible as the theory only compares the relation of proteins of different size, diffusing through the same tissue structures at the same time, usually characterized as diffusion "constants". It is reasonable to suggest that the ratio of these diffusion coefficients (e.g. D_{IgG}/D_{Alb}) does not change in spite of different absolute values of the single diffusion coefficient in different tissue structures;

3. proteins diffuse from blood through the tissue enter the cerebrospinal fluid, and are eliminated by the flowing CSF along its way in the subarachnoid space. The theory does not need to state a CSF flow rate unique for the whole subarachnoid space (laminar vs turbulent flow, etc.).

With these boundary conditions we find an implicit solution for the differential equation of Fick's second law of diffusion [4]. As a consequence, we get a non-linear concentration gradient between blood and CSF in the idealized model. The corresponding sigmoidal curves are shown in Fig. 4 (*left*). The CSF concentrations (Q_A, Q_B, Q_C) of molecules A, B or C depend on the local concentration gradients dc/dx from the surface to subarachnoid space. As the most relevant result of this mathematical treatment, we find that the ratio of CSF concentrations of two different mol-

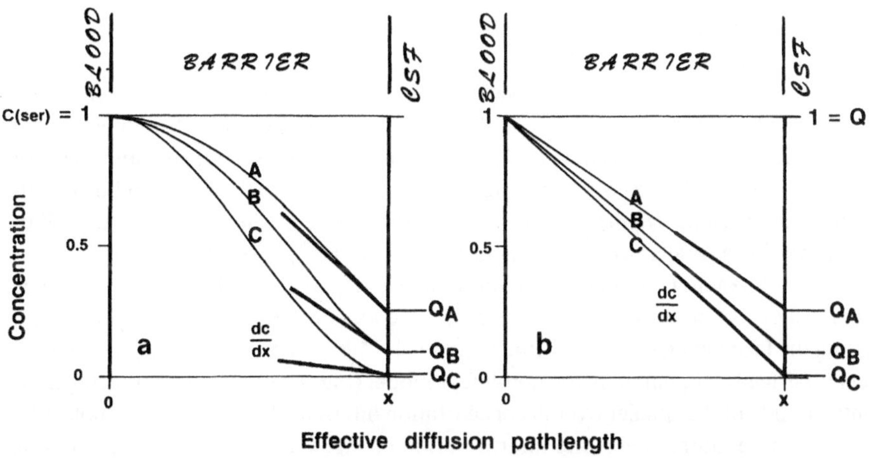

Fig. 4. Comparison of non-linear (*left*) and linear (*right*) models of blood CSF protein concentration gradients. The blood protein concentration of a single protein is normalized with values C(ser) = 1. Corresponding CSF protein concentrations are given as dimensionless CSF/serum quotients (Q_A, Q_B, Q_C) with values between 0 and 1. *x*, represents the effective diffusion pathlength of the idealized barrier according to the boundary conditions described in the text. A, B, C can either represent the curves for molecules of different size (A < B < C) or, alternatively, the concentration distribution of the same protein (e.g. IgM) at different times: before (t_0, curve C) and after (t_1, t_2, curve B and A) the onset of a disease with decreasing CSF flow rate and, subsequently, increasing concentration of IgM in CSF. The local concentration gradient, dc/dx, at border to subarachnoid space (at *x*) and correspondingly the molecular flux increase in model (*left*) from C to A, but decrease in linear model (*right*) from C to A

ecules follows the hyperbolic function, the equation for albumin and IgG shown in the following:

$$Q_{IgG} = \frac{erfc\ z \cdot \sqrt{D_{IgG} / D_{Alb}}}{erfc\ z} \cdot Q_{Alb}$$

The ratio of the both quotients ($Q_{IgG} : Q_{Alb}$) depends only on the diffusion constants of both molecules (D_{IgG} and D_{Alb}) for a certain diffusion path, *z*; *erfc z* is the error function complement of *z*, which can be expressed implicitly as a trigonometrical series for *z*; *z* is a measure of the effective diffusion pathlength between blood and CSF.

The above equation can be written also with the usual hyperbolic formula:

$$Q_{IgG} = a/b \sqrt{Q(_{Alb})_2 + b^2} - c$$

The parameters *a/b*, b^2, and *c* are determined by empirical fit of measured CSF data [4].

From Fig. 4 we learn the difference between the earlier approach of a linear (*right*) concentration gradient and a nonlinear (*left*) concentration gradient between blood and CSF. The basic differences are obtained from the local concentration gradients dc/dx the surface to subarachnoid space. As shown already in Fig. 1 (insertion), the molecular flux, J, of molecules into CSF depends on this concentration gradient dc/dx, too.

In Fig. 4 we have three curves which can be interpreted either as the concentration gradients of different sized molecules (A < B < C), or as the concentration gradients of the same protein species (e.g. IgM) at different times, before (C) and after onset of disease (B, A), with decreasing CSF flow rate (C > B > A). In the model of linear concentration gradients (*right*) it should be stated that the smaller molecule (e.g. albumin) showing a smaller overall gradient (like A, on the *right*) should have the slower molecular flux into CSF than the larger molecule (e.g. IgG, like B on the *right*). This, obviously, must be contradictory to the observed kinetic data (Table 1 and [4]) indicating the faster appearance of serum albumin in CSF than serum IgG in CSF. This effect is perfectly simulated by the non-linear model (Fig. 4, *left*) with larger local gradients (in spite of the smaller overall concentration difference) for the smaller molecule.

Still more interesting is the interpretation of Fig. 4 as a sequence of conditions with reduced CSF flow rate. In both non-linear and linear models, a reduced CSF flow rate with decreasing CSF volume exchange must lead to increasing CSF protein concentrations (Q_C via Q_B to Q_A). But, with varying increases of different protein concentration in CSF controversary consequences will be induced in the different models: if we regard the molecule C with normal CSF flow in the non-linear model, the concentration curve could look like curve C (*left*). With decreasing CSF flow rate, i.e. an increased protein concentration in CSF, the "mean displacement" [4] of diffusing particles, which describes the shift of diffusing molecules into the matter, increases and so we get the new concentration curve shown as curve B and some time later as curve A (*left*). So, secondarily, after initial increase in protein concentration, there is an increase in tissue concentration and, more important, there is an increase of the local concentration gradient (dc/dx) at the surface of the subarachnoid space, i.e. an increased net molecular flux into CSF. In case that Q > 0.5 in the non-linear model (*left*), net molecular flux into CSF would decrease again. This initially non-linear increase of net molecular flux with decreasing CSF flow rate is the basic new statement derived from this model. In contrast, as shown in the linear concentration gradient on the *right* (the older model), whith increasing protein concentration in CSF we have to suggest a decrease of both, local concentration gradient (dc/dx) and net molecular flux into CSF, with increasing barrier dysfunction. This, again, is in contradiction to the observed increase of protein diffusion into CSF in the course of disease.

Thus, the change in CSF flow rate which modulates the protein concentrations in CSF is sufficient to describe quantitatively the empirical data, with increased molecular flux without any change in selectivity or permeability constants [4], i.e. in absence of any changes of the morphological structures relevant for the blood-CSF barrier function.

The rostro-caudal concentration gradient between ventricular and lumbar CSF

The ventricular CSF contains both brain-derived and blood-derived proteins. The concentration of the blood-derived proteins increase steadily in CSF flowing along the neuraxis. Due to the steep blood/CSF concentration gradient of the serum proteins, the molecular flux into CSF exceeds the negligible reflux out of CSF. In Fig. 5, these aspects are shown for IgG and albumin in ventricular, cisternal and lumbar CSF. In case of only blood-derived CSF fractions of IgG and albumin, the IgG quotients follow the hyperbolic discrimination line with increasing albumin quotient between ventricular and lumbar CSF. How can this be understood?

The hyperbolic function, which describes the concentration ratio of two protein species of different size, was derived from conditions at a fixed position in subarachnoid space by varying CSF flow rate (Fig. 1) [4]. From a mathematical point of view there is no difference whether the CSF concentration of a molecule increases as a consequence of reduced CSF volume exchange (slower CSF flow), or due to the steady flux into CSF along the neuraxis (constant CSF flow rate). The common aspect is the modification of the molecular flux J~dc/dx with increase of CSF concentration.

The consequence of this statement, a faster molecular flux into CSF in the lumbar space than in ventricles, was confirmed by experimental data obtained already 30 years ago (cited in [4]). So again, we have two interacting events in CSF along the neuraxis - primarily increasing CSF protein concentration, and secondary increased molecular flux. These aspects are the reason why the same hyperbolic function (e.g. Q_{IgG} = f(QAlb) can be applied to describe blood-derived protein concentrations in ventricular, cisternal and lumbar cerebrospinal fluid, shown for a fictious patient in Fig. 5; the second case of Fig. 5 regards a patient with an additional strong intrathecal IgG synthesis. The steady intrathecal, local synthesis contributes with IgG_{Loc} = 36 mg/L to total CSF IgG concentration in ventricular, cisternal and lumbar CSF. With an increasing blood-derived IgG concentration along the neuraxis, the intrathecal fraction IgG is reduced from 80% (Q_{Alb} = 1.8) to 71% in cisternal CSF (Q_{Alb} = 2.9), and to 60% in lumbar CSF ($Q_{Alb} = 4.5 \cdot 10^{-3}$). As it is not possible to get simultaneously ventricular, cisternal and lumbar CSF from a single patient, the data in Fig. 5 are calculated quotients based on statistical data from a group of patients [11]. These cases in Fig. 5 regarding the rostro-caudal concentration gradients, but with constant CSF turnover, are completely different from the case of the patient with neuroborreliosis (Fig. 7 and Table 3). The changes in the albumin quotients refer to lumbar CSF and indicate a change in CSF flow rate during the course of the disease. In this

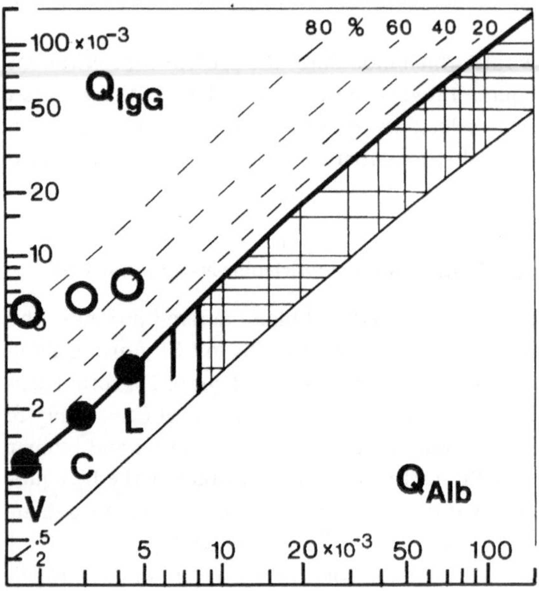

Fig. 5. Protein data in ventricular, cisternal and lumbar CSF in the IgG CSF/serum quotient diagram. The theoretical data of a representative patient (•) with normal albumin and IgG quotients in ventricular (V), cisternal (C) and lumbar (L) CSF are calculated according to [11]. The albumin quotients increase from ventricular CSF to cisternal CSF (1.6 fold) and to lumbar CSF (2.5 fold). For a second fictious patient with a constant intrathecal IgG synthesis (o) of about 18 mg/day or 18 mg/500 mL CSF, we calculate $IgG_{Loc} = 36$ mg/L corresponding to an intrathecal IgG fraction, $IgG_{IF} = 80\%$ in ventricular CSF, decreasing along the neuraxis with increasing blood-derived IgG fraction

Table 3. Variation of blood-derived and brain-derived IgM fractions in CSF with changing CSF flow rate (blood CSF barrier function)

Lumbar Puncture* No.	2	3	4	5
$Q_{Alb} \cdot 10^3$	80	68	26	20
$Q_{IgM} \cdot 10^3$	118	103	30	20
CSF-IgM (mg/L)	141	123.4	36.4	24
IgM_{Loc} (mg/L)	85	77	22.2	14.2
IgM_{IF} (%)	60	62	61	59

* Lumbar punctures of the same patient with neuroborreliosis shown in Fig. 8 at 6, 8, 10 and 16 weeks after tick bite. These data represent the time of almost constant intrathecal IgM fraction during recovery of the patient with decreasing albumin quotient due to increasing CSF flow rate (CSF turnover)

case a constant intrathecal IgM fraction (IgM_{IF} = 60%) lasted over several weeks. As shown in Table 3, during the recovery of the patient with decreasing albumin quotient, the total CSF-IgM concentration as well as the brain-derived IgM_{LOC} decrease with a constant ratio indicated by the intrathecal fraction IgM_{IF} = 59 -61%.

The reduced CSF turnover observed in many neurological diseases influences the brain-derived protein fractions (e.g. due to intrathecal IgM synthesis) in CSF as well as the blood-derived CSF fraction. Both changes are a consequence of the reduced CSF volume exchange per unit of time. With a constant amount of protein released from brain per day, the CSF concentration increases with decreasing CSF turnover.

The new paradigm of blood CSF barrier function and dysfunction - Biophysical model

1. Application of the molecular flux/CSF flow model with relevant boundary conditions lead to a non-linear concentration gradient at the barrier between blood and CSF. (The first derivative of this sigmoidal tissue concentration curve is a Gaussian error curve).
2. The concentration curve is characterized by the mean molecular displacement or penetration depth. The mean molecular displacement [4] is generally a function of molecular size, and a function of CSF flow rate in this particular model.
3. Decreasing of CSF flow rate induces: a) reduced volume exchange, with a subsequent increase of protein concentration in CSF; b) an increasing mean displacement with nonlinear increase of the molecular flux, J.
4. The concentration ratio of two molecules in CSF with different size (e.g. Q_{IgG}: Q_{Alb}) changes with increasing protein concentration in CSF, i.e. with a decrease of CSF flow rate according to the following hyperbolic function:

$$Q_{IgG} = \frac{erfc\ z \cdot \sqrt{D_{IgG} / D_{Alb}}}{erfc\ z} \cdot Q_{Alb}$$

This function, which shows that the ratio of Q_{IgG}: Q_{Alb} depends only on the diffusion constants D_{IgG}/D_{Alb}, is valid in ventricular, cisternal or lumbar CSF space.
5. The steady diffusion of molecules into CSF along the neuraxis lead to a rostro-caudal concentration gradient. This gradient is non-linear and explains the observed faster molecular flux into lumbar CSF compared to ventricular CSF (again, caused by increasing mean displacement with increasing of molecular flux into CSF).

Physiological and pathophysiological interpretations

With the actual CSF flow model many unexplained facts from the physiology of CSF and pathophysiology of neurological diseases could be described in a new light [4]. The normal human newborn has extremely high CSF protein concentrations with Q_{Alb} values up to $30 \cdot 10^{-3}$. Earlier interpretations suggested an immature blood CSF barrier, with poor selectivity and hight permeability. However, there is no doubt that early in gestation the anatomical structures for barriers are present [12]. In the quotient diagrams there is no difference in selectivity, i.e. in molecular size-dependent discrimination by the blood CSF barrier function for proteins in newborn children, when compared to mature adults with corresponding albumin quotients. The reported prenatal onset of CSF flow [12] must gradually reduce the protein concentration in CSF as the flow rate increases. It might be due to the late structural changes of the arachnoid villi and granulations, which bring CSF drainage to a maximum about 4 months after birth with a minimum of Q_{Alb} at this time.

In the mature human a continuously increasing CSF protein concentration is observed. This age-dependent increase could be explained as a decrease in CSF production rate in elderly volunteers [13], with a subsequently reduced CSF flow rate (0.4 mL/min in young to 0.1 mL/min in elderly volunteers).

For CNS leukemia, which is primarily an arachnoid disease with changes in trabeculae, a reduced CSF flow was suggested from histopathology. A purulent bacterial meningitis is accompanied by an increased CSF viscosity and meningeal adhesions. Protein complexes and cell deposits in the arachnoid villi have been detected in post mortem material. This, again, represents a severe handicap for CSF bulk flow.

Polyradiculitis of the type Guillain-Barré may be accompanied by swelling in the region of the spinal roots (cauda equina), probably reducing flow through arachnoid villi into the veins associated with spinal nerve roots. In spinal blockade (Froin's syndrome), caudally to the blockade, high serum protein values are measured in lumbar CSF in spite of normal cisternal or ventricular CSF values. In contrast to blood-derived proteins, the brain-derived proteins (prealbumin) have a decreased concentration relative to albumin caudal to a spinal blockade. In this case, again, the molecular size-dependent discrimination (selectivity) for protein transfer between blood and CSF is not disturbed.

Ascorbate, a low molecular weight substance, is 6 fold increased in human lumbar CSF over blood concentrations due to an active transport through the choroid plexus. The decrease of ascorbate concentrations in the blood; in cases of blood-CSF barrier dysfunction, can be explained by the decreasing CSF flow rate [9]. This decrease of blood ascorbate concentration contradicts a barrier "leakage", by which an increased instead of a decreased ascorbate concentration in blood should be induced. This extraordinary example from a small molecule in the CSF offers another strong argument against changes in structures in case of blood CSF barrier dysfunction.

A larger mean IgG-Index ($Q_{IgG}/Q_{Alb} = 0.8$) in smaller mammalian species (rats, guinea pigs [14]) compared to an index of 0.43 in humans [11] can be partly explained by a shorter effective pathlength for protein diffusion from blood to CSF, and mainly by a difference in CSF flow rate due to different CSF production rates [1]. There is no necessity to suggest a difference in selectivity of the barrier function. The increased CSF protein content in experimental allergic encephalomyelitis of guinea pigs [14], regarded as model for MS, might be explained by its primarily spinal lesions, which is different from the disseminated process in MS, with only minor changes in CSF protein content.

The CSF/serum quotient diagrams

The CSF/serum quotient diagrams, improved according to Reiber [4], are shown in Fig. 6. They indicate the hyperbolic lines (bold lines) for discrimination between blood-derived CSF fractions. Compared with earlier diagrams [15], the sensitivity

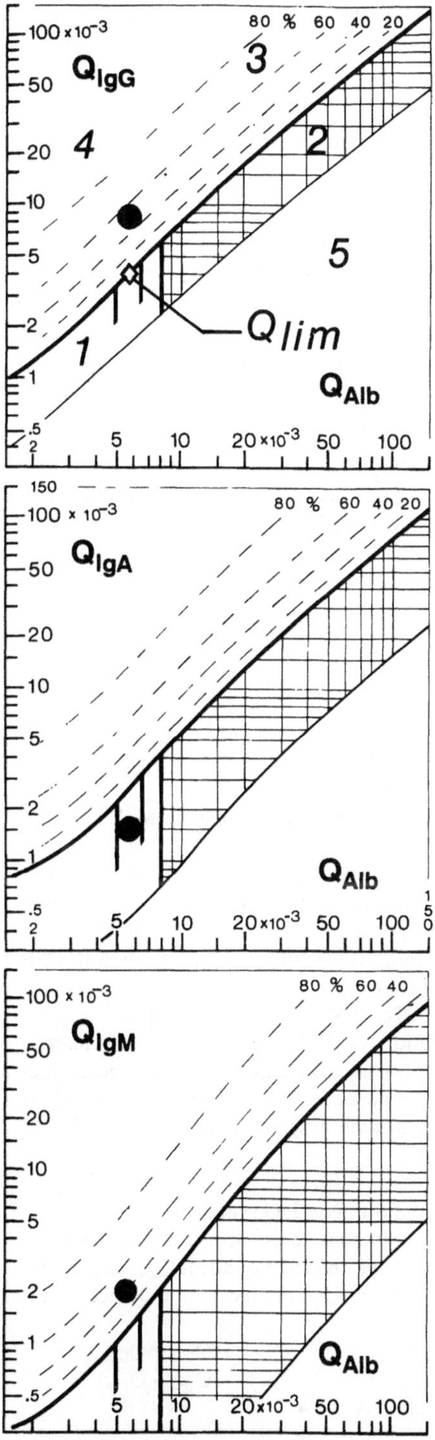

Fig. 6. CSF/serum quotient diagrams for IgG, IgA, IgM [10]. The hyperbolic curves (*thick lines*) represent the discrimination lines between brain-derived and blood-derived Ig fractions. Values above these upper discrimination lines represent intrathecal IgG, IgA or IgM synthesis. The *dashed lines* indicate the extent of the intrathecal synthesis as intrathecal fractions (IF) in percent of total CSF concentration of the immunoglobulin. The limit of reference range between normal and increased CSF protein concentration with reduced CSF turnover (blood CSF barrier dysfunction) is indicated by the age-dependent vertical lines at $Q_{Alb} = 5\times10^{-3}$ [up to 15 years]; at $Q_{Alb} = 6.5\times10^{-3}$ [up to 40 years]; at $Q_{Alb} = 8\times10^{-3}$ [up to 60 years]. The diagrams depict five ranges: **1** = normal; **2** = reduced CSF turnover i.e. blood CSF barrier dysfunction; **4** = intrathecal Ig synthesis without change in CSF turnover; and **3** = intrathecal Ig synthesis with a reduced CSF turnover, i.e. increased Q_{Alb}. Values below the lower hyperbolic line in range **5** indicate a methodological fault. The data given in the diagrams are from a multiple sclerosis patient who showed a two-class Ig response with the intrathecal fractions $IgG_{IF} = 55\%$, $IgA_{IF} = 0\%$, and $IgM_{IF} = 40\%$. For calculation of Antibody-Index results, the specific CSF/serum quotient refers to Q_{Lim}, the value of the hyperbolic discrimination function for the individual albumin quotient of this patient

for detection of intrathecal humoral immune response in lumbar CSF of children or in ventricular CSF is increased, and the logarithmic scales are used with an extended range up to $Q_{Alb} = 150 \cdot 10^{-3}$. The hyperbolic functions are valid in the whole biological range up to the largest albumin quotients measured so far ($700 \cdot 10^{-3}$).

Therefore, patients with results outside the diagrams can be characterized numerically by the intrathecally (locally) synthesized IgG, IgA and IgM fraction as Ig_{Loc} in mg/L, or Ig_{IF} in % of total CSF concentration. Due to the analytical imprecision (CVs of 3 - 8%) for albumin, IgG, IgA, and IgM quotients (Q_{Alb}, Q_{IgG}, Q_{IgA}, Q_{IgM}), intrathecal synthesis is resported only if the intrathecal fraction Ig_{IF} is greater than 10%.

In addition to the upper discrimination line, these improved diagrams also show the lower border lines of the reference range which, between upper and lower border lines, includes 99% of the 4300 cases investigated to date [4]. As mentioned above, the diagrams established for lumbar CSF can be applied to ventricular and cisternal CSF for detection of intrathecal fraction of IgG, IgA or IgM in CSF without any restriction or correction. Of course, due to the concentration gradient between ventricular and lumbar CSF, different reference ranges [10] for detection of a blood CSF barrier dysfunction (Q_{Alb}) should be used.

Numerical evaluation of CSF protein data

1. The general hyperbolic function $QIg = a/b \sqrt{Q(Alb)^2 + b^2} - c$ has the following equations to describe the upper limit Q_{Lim} (Ig) of the reference range in the CSF/serum quotient diagrams:

$$Q_{Lim}(IgG) = 0.93 \sqrt{(QAlb)^2 + 6 \cdot 10^{-6}} - 1.7 \cdot 10^{-3}$$
$$Q_{Lim}(IgA) = 0.77 \sqrt{(QAlb)^2 + 23 \cdot 10^{-6}} - 3.1 \cdot 10^{-3}$$
$$Q_{Lim}(IgM) = 0.67 \sqrt{(QAlb)^2 + 120 \cdot 10^{-6}} - 7.1 \cdot 10^{-3}$$

Values for Q_{IgG}, Q_{IgA}, Q_{IgM} above this hyperbolic discrimination line indicate an intrathecal synthesis.

2. Quantitation of intrathecal synthesis. The amount of locally synthesized immunoglobulins released into CSF can be expressed either as contribution to the CSF Ig concentration:

$$Ig_{Loc} = [Q_{Ig} - Q_{Lim}(Ig)] \cdot Ig_{serum} \qquad [mg/L]$$

or as the intrathecal fraction, Ig_{IF}, referring Ig_{Loc} to the total Ig concentration in CSF (Ig_{Loc}/Ig_{CSF}), and rearranged with $Q_{Ig} = Ig_{CSF}/Ig_{serum}$:

$$Ig_{IF} = [1 - Q_{Lim}(Ig) / QIg] \cdot 100 \qquad [\%]$$

The locally in CNS synthesized IgG concentration (i.e. IgG_{Loc} in mg/L) depends numerically on the blood CSF barrier function (CSF flow rate). In contrast, the intrathecal fraction (i.e. IgG_{IF} in % of total CSF IgG) remains independent of the CSF flow rate and offers the better term to determine dominance of intrathecal synthesis amongst the different immunoglobulin classes ($IgG_{IF}/IgA_{IF}/IgM_{IF}$).

3. Dominance amongst intrathecal fractions:

$IgG_{IF} > IgM_{IF}$ means dominant intrathecal IgG synthesis

4. Antibody-Index [16]. The specific intrathecal immune response of a certain antibody species is calculated with the specific CSF/serum concentration quotient Q_{spec} and the total IgG (or IgM class) quotient either as the empirical Q_{IgG} or, in case of polyspecific immune response, as $Q_{Lim}(IgG)$:

$$AI = Q_{spec} / Q_{IgG} \qquad (Q_{IgG} < Q_{Lim})$$
$$AI = Q_{spec} / Q_{Lim}(IgG) \qquad (Q_{IgG} > Q_{Lim})$$

Reference range AI = 0.7 –´1.3; Pathological values AI > 1.4

Calculation example for quantification of intrathecal immune response

The points in Fig. 6 represent quotients from a multiple sclerosis patient with the following data: Alb(CSF) = 259 mg/L, Alb(serum) = 44.6 g/L; IgG(CSF) = 68.9 mg/L; IgG(serum) = 8.1 g/L; IgA(CSF) = 2.25 mg/L; IgA(serum) = 1.5 g/L; IgM(CSF) = 1.9 mg/L; IgM(serum) = 0.95 g/L; $Q_{Alb} = 5.8 \cdot 10^{-3}$; $Q_{IgA} = 1.5 \cdot 10^{-3}$; $Q_{IgM} = 2.0 \cdot 10^{-3}$. For calculation of intrathecal synthesis and Al values we need: $Q_{Lim}(IgG) = 4.6 \cdot 10^{-3}$, $Q_{Lim}(IgA) = 2.7 \cdot 10^{-3}$, and $Q_{Lim}(IgM) = 1.2 \cdot 10^{-3}$. In this patient the intrathecal fractions (IF) are: $IgG_{IF} = 54.6\%$, $IgA_{IF} = 0\%$, $IgM_{IF} = 40\%$. They indicate synthesis of two immunoglobulin classes in the CNS with dominance of IgG_{IF}.

For the calculation of the antibody index values with the specific quotients (IgG class) for measles, Q (measles) = $26.6 \cdot 10^{-3}$, for rubella Q(rubella) = $22.0 \cdot 10^{-3}$, and for varicella zoster antibodies Q(VZV) = $5.0 \cdot 10^{-3}$, we get with $Q_{Lim}(IgG)$ the following specific antibody-index values: measles-AI = 6.4, rubella-AI = 5.3, and VZV-AI = 1.2. These results for the polyspecific intrathecal immune response indicate intrathecal measles and rubella antibody synthesis – a combination highly suggestive for a chronic inflammatory disease (autoimmune type).

Clinical relevance

Neuroimmunological reactions

The humoral immune response in CNS is different from the immune response usually observed in blood. As a main difference we find no switch from IgM class response to a more specific IgG class response in the course of inflammatory diseases. The pattern of intrathecal IgG/IgA/IgM synthesis remains rather constant and depends on the cause, pathophysiology and localization of the disease process.

As a second difference in neuroimmunological processes, we find a slow, longlasting decay of intrathecal antibody synthesis, sometimes detectable 10-15 years after sufficient treatment (neurosyphilis, neuroborreliosis, or HSV-encephalitis).

Both aspects, the lack of IgM to IgG switch and the slow normalization of intrathecal antibody synthesis, could be the consequence of the same problem: a handicapped regulation of intrathecal immune response. Due to the barrier-dependent low immunoglobulin concentration in CNS, and the only local (perivascular) invasion of relatively few immunocompetent cells, we might calculate a 10^5 lower probability for an encounter of cells and antibodies, compared to blood. Irrespective of the discontinuous distribution of immunocompetent cells, this figure is suggestive of a compartmental regulation problem. From a diagnostic point of view, the lack of IgM to IgG switch in CNS is the chance to characterize disease-related instead of acuity-related patterns. However, this is also a diagnostic disadvantage: the characterization of the acuity of disease (see below).

Disease-related patterns in quotient diagrams

The following examples are somehow representative cases:

- Figure 7, Case 1. In the absence of a humoral immune response, the IgG, IgA and IgM quotient follow the hyperbolic function, in the diagrams shown for a benign bacterial meningitis.
- Figure 7, Case 2. There is no switch from IgM class to IgG class response in the central nervous system. The initially dominant IgM synthesis remains dominant over weeks, as shown for the neuroborreliosis (see also Fig. 10). Such a dominant humoral, together with cellular IgM response associated with a blood CSF barrier dysfunction, indicates a neuroborreliosis with such a high specificity as 96% and a sensitivity of 70% without a borrelia-specific analysis [17].
- Figure 8. In neurotuberculosis, we find frequently an isolated or dominant IgA synthesis together with a blood CSF barrier dysfunction. The combination with increased CSF lactate and an intermediate pleocytosis is highly plausible to suggest this diagnosis. Only in cases of *Spondylitis tuberculosa* with a spinal location of the process an additional IgM synthesis has been observed.
- Figure 9. In neurosyphilis we find different patterns for different pathomechanisms. Both the meningovascular and the parenchymatous type of disease show almost normal CSF flow rate associated with intrathecal IgG synthesis. But the extremely strong IgM synthesis, dominant over IgG synthesis in the parenchymatous progressive paralysis, is not seen in the meningovascular type of disease. In both cases the absence of an IgA synthesis is a characteristic feature. For diagnosis, the specific antibody response against *Treponema pallidum* remains, of course, the most important information.
- Figure 10. Patterns of immunoglobulin response in the juvenile immune system are demonstrated to be similar like that in adults. Only the smaller albumin quotients, due to a faster CSF flow rate and shorter CSF flow distance in this age, are different.

Intrathecal IgM and IgA synthesis

The intrathecal IgM synthesis is not a sign of acute disease as shown in Fig. 6, 7 and 9. The intrathecal IgM class response seen in progressive paralysis, tuberculous spondylitis, neuroborreliosis, mumps meningitis, and multiple sclerosis with different frequencies and intensities offers less information on its own. It is in combination with other CSF data that IgM contributes to a more or less typical pattern.

This is in contrast with intrathecal IgA response at the time of first diagnostic puncture, which typically indicates a bacterial origin of the disease; however, this response is, again, of different frequency for different bacteria, like *Neisseria meningitidis* or *Staphylococcus*, but in later course of viral infections intrathecal IgA synthesis combined with IgM synthesis is observed too, e.g. in herpes simplex encephalitis. Apparently, the intrathecal immune response is more related to the special pathomechanisms of the diseases with its different time courses and locations.

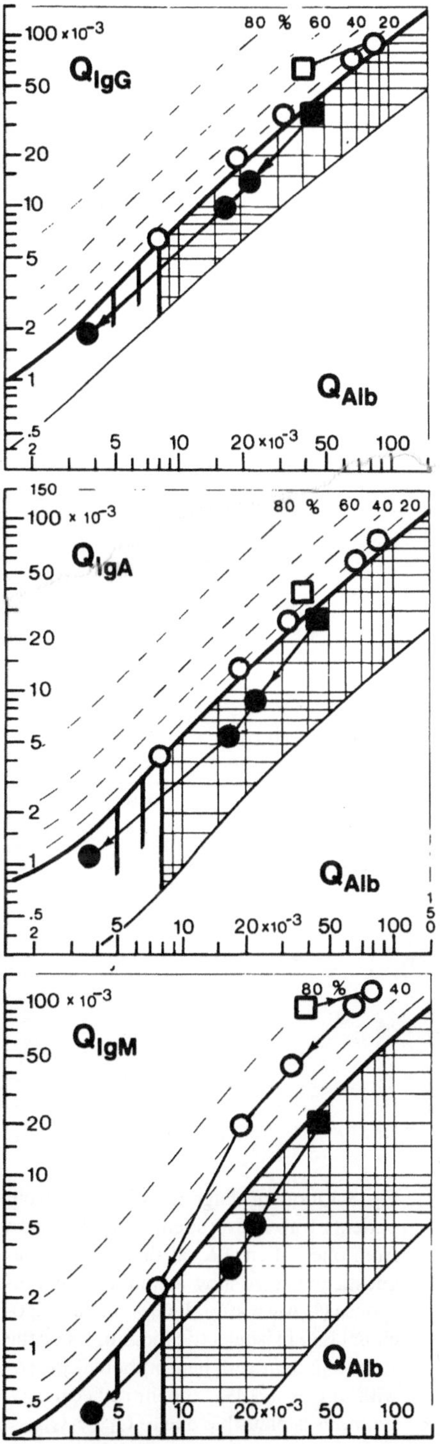

Fig. 7. The time course of CSF protein data from a patient with a meningococcal meningitis (•), and from a patient with a neuroborreliosis (o). From the patient with a meningococcus meningitis CSF was obtained at the days 1 (■), 3, 6 and 13. Cell counts were 7250/µL, 2730/µL and 2/µL, respectively. In the patient with neuroborreliosis CSF was obtained at 3 (□), 4, 6, 10, 16 and 83 weeks after tick bite with the sequence of data indicated in the IgM diagram; cell counts were 132/µL, 100/µL, 39/µL, 90/µL, 15/µL and 3/µL, respectively. ■, □ represent results from first diagnostic samples

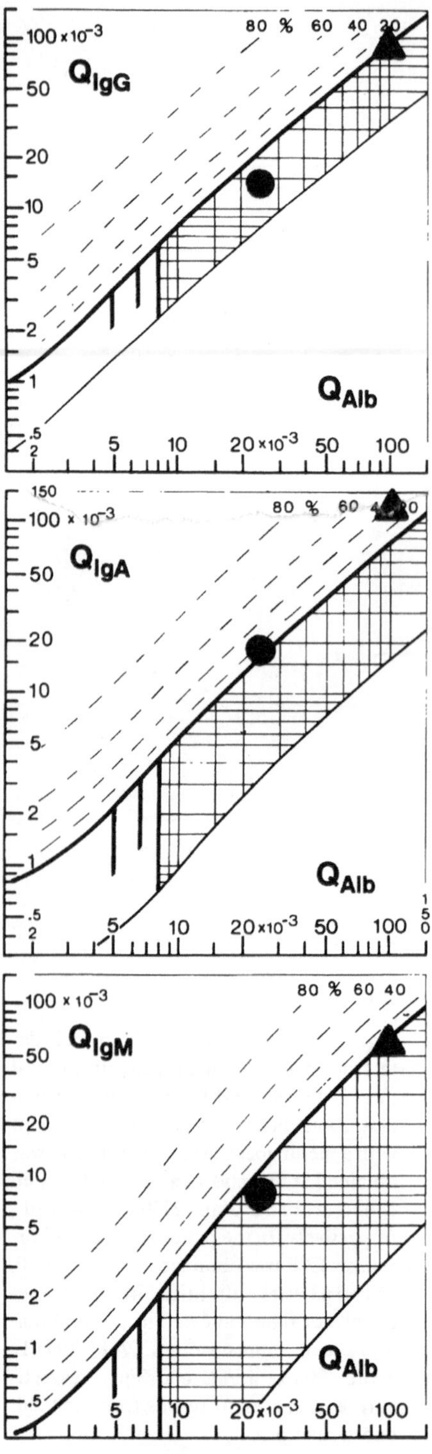

Fig. 8. Quotient diagrams with data from patients with neurotuberculosis. The data from one patient (▲) at time of first diagnostic puncture show a dominant IgA response ($IgA_{IF} = 40\%$, $IgG_{IF} = 0$, but oligoclonal IgG in isoelectric focusing). For the second patient with neurotuberculosis (•) with $IgG_{IF} < 10\%$, the intrathecal IgA synthesis is plausible due to $Q_{IgA} > Q_{IgG}$

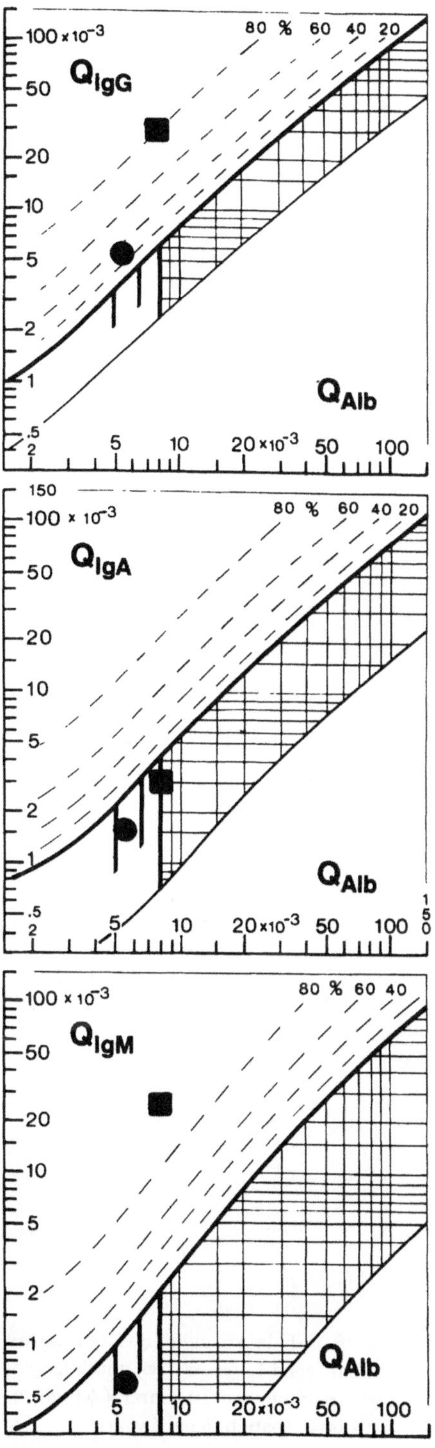

Fig. 9. Quotient diagram from patients with neurosyphilis. The two patients are representative for the meningovascular type (•) with an intrathecal synthesis ($IgG_{IF} = 30\%$), and for progressive paralysis, parenchymateous form [■]), with a dominant intrathecal IgM synthesis ($IgM_{IF} > 80\%$) in addition to a strong intrathecal IgG synthesis ($IgG_{IF} = 80\%$). In both cases no intrathecal synthesis of IgA can be observed

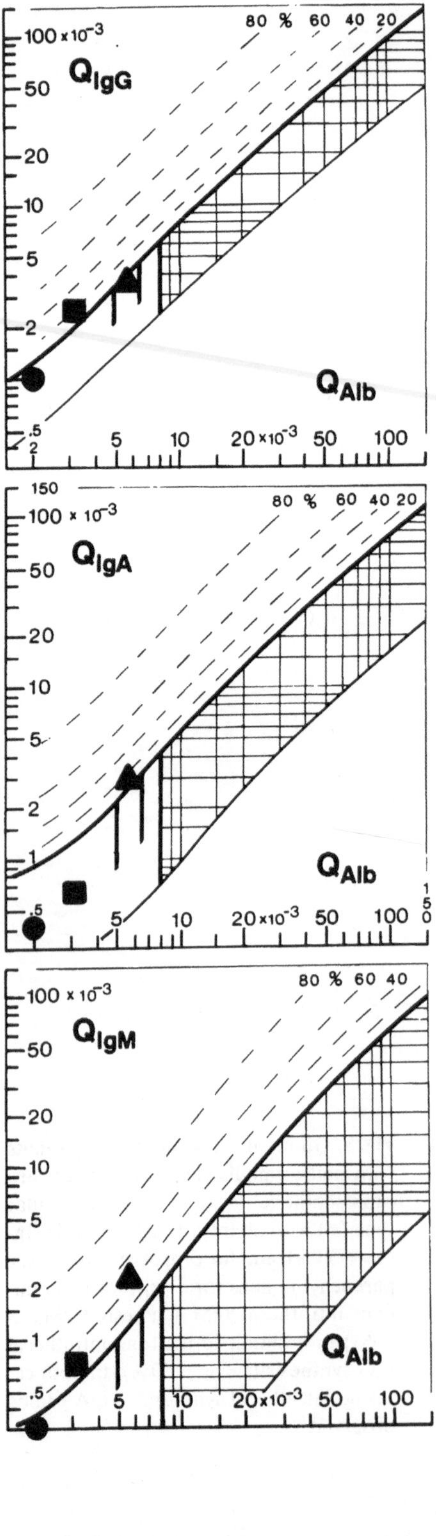

Fig. 10. CSF data in childhood. (●) control patient (aged 1.5 years); (■) multiple sclerosis (patient aged 10 years); (▲) neuroborreliosis (patient aged 5 years)

Signs of acute disease in CNS

The main signs in cerebrospinal fluid of an acute, active disease in CNS are the increased CSF cell count and an increased Q_{Alb}, i.e. reduced CSF turnover. The presence or absence of an intrathecal humoral immune response cannot be used as a sign of disease duration. There are always three different possible interpretations of intrathecal antibody synthesis:

1. acute disease with reaction to a monospecific antigen;
2. postacute, delayed, decreasing antibody synthesis without any clinical relevance;
3. part of a polyspecific immune response typical for acute and in particular chronic inflammatory diseases.

Relevance of CSF analysis

The combined analysis of IgG, IgA and IgM in CSF, together with CSF turnover (Q_{Alb}) is now widely recognized as a helpful tool for differential diagnosis of neurological diseases. In addition, of course, there is no case in which CSF analysis can be used as the only source of informations for a secure diagnosis. CSF analysis cannot now, and never will replace clinical informations. However, in the frame of incisive, clinically well-founded questions (to be reported to the laboratory), CSF protein analysis offers relevant information for confirmation of, or discrimination among differential diagnostic alternatives.

In addition, and independently of the quotient patterns, IgG and IgM quotients are needed for calculation of a most sensitive detection of intrathecal synthesis of specific antibodies for diagnosis of certain diseases, e.g. VZV induced infections. Long experience indicates that CSF data are helpful to the physician in direct proportion to the physician's clinical expertise.

Acknowledgements. I thank Dr. Petra C. Schlüter for the very helpful comments which improved this contribution.

References

1. Davson H, Welch K, Segal MB (1987) Physiology and pathophysiology of the cerebrospinal fluid. Churchill, London
2. Wood JH (1980 and 1983) Neurobiology of cerebrospinal fluid, vols 1 and 2. Plenum Press, New York
3. Thompson EJ (1988) The CSF proteins: a biochemical approach. Elsevier, Amsterdam
4. Reiber H (1994) Flow rate of cerebrospinal fluid (CSF) - a concept common to normal blood-CSF barrier function and to dysfunction in neurological diseases. J Neurol Sci 122: 189-203
5. Reiber H (1994) The hyperbolic function: a mathematical solution of the protein flux/CSF flow model for blood-CSF barrier function. J Neurol Sci 126: 243-245
6. Reiber H (1995) Biophysics of protein diffusion from blood into CSF: the modulation by CSF flow rate. In: Greenwood J, Begley D, Segal M (eds) New concepts of a blood-brain barrier. Plenum, London, pp 219-227

7. Bradbury M (1979) The concept of a blood-brain barrier. John Wiley and Sons, Chichester
8. Kruse T, Reiber H, Neuhoff V (1985) Amino acid transport across the human blood-CSF barrier. An evaluation graph for amino acid concentrations in cerebrospinal fluid. J Neurol Sci 70: 129-138
9. Reiber H, Ruff M, Uhr M (1993) Ascorbate concentration in human cerebrospinal fluid (CSF) and serum. Intrathecal accumulation and CSF flow rate. Clin Chim Acta 217: 163-173
10. Reiber H (1995) External quality assessment in clinical neurochemistry: survey of analysis for cerebrospinal fluid (CSF) proteins based on CSF/serum quotients. Clin Chem 41: 256-263
11. Reiber H, Thiele P (1983) Species-dependent variables in blood cerebrospinal fluid function for proteins. J Clin Chem Clin Biochem 21: 199-202
12. Mollgard K, Saunders NR (1966) The development of the human blood-brain and blood-CSF barriers. Neuropath Appl Neurobiol 12:337-358
13. May C, Kaye JA, Atack JR, Schapiro MB, Friedland RP, Rapoport SI (1990) Cerebrospinal fluid production is reduced in healthy aging. Neurol 40: 500-503
14. Suckling AJ, Reiber H, Rumsby MG (1986) The blood-CSF barrier in chronic relapsing experimental allergic encephalomyelitis. In: Suckling AJ, Rumbsby MG, Bradbury MWB (eds) The Blood-Brain-Barrier in Health and Disease. Ellis Horwood, Chichester, pp 147-157
15. Reiber H, Felgenhauer K (1987) Protein transfer at the blood cerebrospinal fluid barrier and the quantitation of the humoral immune response within the central nervous system. Clin Chim Acta 163: 319-328
16. Reiber H, Lange P (1991) Quantification of Virus-Specific Antibodies in Cerebrospinal fluid and Serum: Sensitive and Specific Detection of Antibody Synthesis in Brain. Clin Chem 37: 1153-1160
17. Tumani H, Nölker G, Reiber H (1995) Relevance of cerebrospinal fluid parameters for early diagnosis in neuroborreliosis. Neurology 45: 1663-1670
18. Felgenhauer K, Reiber H (1992) The diagnostic relevance of antibody specificity indices in multiple sclerosis and herpes virus induced diseases of the nervous system. Clin Invest 70: 28-37
19. Kahle W, Leonhardt H, Platzer W (eds) (1991) Taschenatlas der anatomie. Vol 3, Thieme-Verlag Stuttgart, pp 262-271

Lymphocyte subsets in multiple sclerosis cerebrospinal fluid

M.G. Marrosu

The beginning of CSF lymphocyte subset study

The study of cerebrospinal fluid (CSF) cells in multiple sclerosis (MS) has been closely linked to progress in the field of general immunology. In the mid-70s, with the advent of the idea that MS was an immunologically-mediated disease (or, at least, that an immunological process was at the basis of its pathogenesis), there was a sudden fluorishing of research on CSF cells. In those years, immunological methods used to distinguish B from T lymphocytes were applied to CSF cells, in order to establish the relative percentage of T and B lymphocyte subsets. The challenge was to find some numerical or functional abnormality to explain the high production of IgG in MS CSF.

Because of the clinical and pathological resemblance of MS to experimental allergic encephalomyelitis (EAE) [1], a T cell-dependent disease, T lymphocytes are believed to play a central role in the disease process of MS, and methodologies used in such research are derived from general immunology. The rosetting technique, in particular, was able to differentiate three types of lymphocytes: the *active* T-cell subpopulation (with very high activity for sheep red blood cells, E rosettes); T cells bearing receptor for IgG Fc (T Fc^+ mainly suppressor cells, EA rosettes), and B cells (EAC rosettes). Retrospectively speaking, the major goal of these studies was to demonstrate that T cells represent the greatest proportion of CSF cells and that activated lymphocytes are present in greater quantities in CSF than in peripheral blood [2, 3]. The ability of lymphocytes to cross the blood-brain barrier (BBB) appears to be in proportion to their degree of activation. The observation that immunological blood phenomena do not reflect what is happening inside the brain is rendered quite obvious by this characteristic. In other words, CSF immunity is not a perfect mirror of blood immunity. As deduced from several reports [2, 8], it is believed that the most important alterations in the immune system in MS reside in CSF rather than in peripheral blood (PB). With regard to MS CSF studies, from 1975 to 1980 a relatively important body of work was produced (reviewed in [7]), often generating contradictory data but enough evidence for immunological dysfunction, particularly regarding the total T subpopulation which appeared to be reduced in MS, mainly during exacerbations [6-8]. An interesting point was raised by the

II Chair of Paediatric Neuropsychiatry, Department of Neuroscience, University of Cagliari, Via Ospedale 119, 09124 Cagliari, Italy

finding of MS CSF T cell reduction [7], suggesting an impaired suppressor effect on B cell proliferation and IgG synthesis, which accounted for the IgG hyperproduction observed in MS CSF.

Cells in MS CSF

In MS, the cell count shows a slight but definite increase above the value of less than 4 cells/mc, considered normal. Nevertheless, this increase is too low to allow an accurate study of specific cell characteristics in terms of both membrane receptor expression and functional status. Since 1980, some strategies have been developed for studying the functional properties of CSF cells in conditions as similar as possible to those occurring *in vivo*. The advent of monoclonal antibodies (mAbs) marked a milestone in methodological approaches to lymphocyte subset studies [9]. Flow cytometry, able to analyze a single cell by a laser beam, made it possible to study one or more CSF lymphocyte membrane receptors, developing a previously unimaginable field of research. As the number of mAbs increased, the possibility to identify a lymphocyte subset responsible for myelin damage, and to characterize CSF cellular response in MS in terms of functional activity became more feasible.

CSF cells determined by monoclonal antibodies

Studies with mAbs reported that T cells (CD3) predominate in normal CSF [10], thus confirming previous reports using the E-rosette formation method. Moreover, T helper/suppressor cell ratio (CD4/CD8) in CSF was the same of that found in PB [10]. Using mAbs, various studies were performed on MS in order to explain the variation in cell subsets during different phases of the disease. The majority of these studies lacked a “normal” control population, a bias which often seems to produce conflicting data and hinder the understanding of results. Only in a very few reports [10-14], CSF cells were studied in both MS and control population. These studies showed that, during relapse, CSF T suppressor cells (CD8) fall [8, 10-14], while a low T suppressor count was found in progressive MS [14], suggesting an ongoing immunological defect in CD8 cells during the progressive phase of the disease. Whether or not CD8 impairment is the cause of both acute MS attacks and progression is difficult to say, but a longitudinal study [14] showed that suppressor cell imbalance is constant over time, suggesting a true role of this subset in relapse and/or progression of the disease. Moreover, corticosteroid therapy performed on MS relapses showed a selective loss of CD4 and, to a lesser extent, of CD8 cells, suggesting that steroids may act mainly on T helper subset [13-15]. The activation state of CSF cells was studied by two different mAbs: one was anti-DR (HLA-DR related antigen), expressed by several immunocompetent cells during the late activation phase [16], and which identifies cell surface glycoprotein coded by genes of the HLA-DR region; the other was anti-IL-2 (interleukin-2 receptor), present only in activated cells and expressed in the early activation phase [17]. Several studies reported that the percentage of IL-2 positive cells is higher in MS CSF [16, 18, 19], suggesting the presence of a constant antigenic stimulus in CSF (myelin, viral, others?), or a selective passage of peripherally activated cells, as demonstrated in EAE [20]. This last hypothesis may be supported by the presence of a high number of PB cells carrying both Ia antigen and Il-2 receptor during the remission phase of MS [21].

Another step in the saga of MS and mAbs was taken with the use of two color fluorescence analysis, simultaneously using phycoerytrin-conjugated and fluorescein-conjugated mAbs by a cell-sorter equipped with a dual laser for green and red fluorescence. This method was able to contemporaneously differentiate the subset of CSF cells having two membrane receptors, providing more functional information on the role of CSF lymphocytes. In particular, studies performed using two mAbs, CD45R (2H4) and CDw29 (4B4), seem to be interesting. According to one interpretation [22], cells bearing CD4 and 2H4 were considered to have a suppressor-inducer capability on Ig secretion, at the opposite of CD4+4B4 cells considered to play a helper-inducer role [23]. Other observations [20, 24, 25] have led to the interpretation of the fundamental difference between 2H4 and 4B4 subsets according to their prior activation by exposure to antigen, so that 2H4 are virgin or naive cells, while 4B4 are memory cells. Studies of CSF cells, stained with both mAbs, produced unclear results. Some Authors reported a decrease in CD4+2H4+ subset in both progressive [26] and acute [27] MS. On the contrary, no variation in this subset was found by others [28, 29], while a reduction in 4B4+ (helper-inducer, memory subset) and an increase in 2H4+ (helper-suppressor, naive subset) was reported in acute MS [29]. It is noteworthy that different methods and sources of mAbs were used in the various studied [26-29]; thus, interpreting the data appears particularly difficult. Moreover, the low number of cells, the difficulty in performing serial studies and, finally, doubts regarding the functional role of 2H4 and 4B4 cells are all strong limitations on understanding the significance of the reported data.

Some interesting work concerned the study of adhesion molecules in CSF cells. Such studies arose from the consideration that specific T cells need to migrate from the bloodstream into CSF compartment in order to initiate the inflammatory reaction in MS, thus expanding to myelin-specific clones. Adhesion molecules are a group of specialized surface molecules, which act on cell contact and migration [30], and mediate the recruitment of lymphocytes in inflammatory reactions. Using two-color fluorescence analysis and a set of mAbs which identified several types of adhesion molecules, like intercellular adhesion molecule (ICAM), very late activation antigen (VLA) and lymphocyte function activated antigen (LFA), a high expression of T memory cells carrying several adhesion- and activation-related molecules has been found in MS, in aseptic meningitis and in normal controls [31]. This supported the idea that T memory cells with a high expression of several adhesion molecules are selectively recruited by the central nervous system compartment in both normal and inflammatory conditions, as part of a general process of immunosurveillance.

CSF lymphocyte subset studies using cultured cells

The study of immunoregulatory CSF cells is hindered by limited quantity of material obtained with a single lumbar puncture. Moreover, since for ethical reasons performing serial analysis without therapeutic justification is forbidden, the majority of reports provided information limited to a given time (generally during a relapse) in the immunological history of the disease. Since the challenge was to obtain enough cells to permit more detailed studies, several investigators expanded CSF T cells in lines and clones using interleukin 2 [32, 33]. However, cell lines do not necessarily reflect the original T cell population, due to possible selection by *in vitro* culture. Aside from this possible selection bias, these studies revealed that the majority of CSF-derived lines were characterized by activated T cells [34] with both helper/inducer and suppressor/cytotoxic phenotype [35], and that more than 90% of CD8-derived CSF cells from MS and controls were cytotoxic precursors, while no cells with natural killer-like activity were identified [36]. Moreover, the percentage of HLA-DR+ cells increased [37], thus demonstrating once again that a selected activated T cell population is present in CSF.

A more sophisticated approach was constituted by the use of a sensitive microculture system, utilizing a mitogenic anti-CD3 mAb and Il-2, which allowed clonal expansion of all T lymphocytes [38]. By using this microculture system, the frequency of cytotoxic and natural killer (NK) T lymphocyte precursors was determined, showing that in MS, as in inflammatory diseases, the relative number of T cells with cytolytic and/or NK potential increased, and that a substantial number of these cells possessed the CD4 phenotype [38].

Conclusions

Despite numerous studies in the field of CSF cells, the role of various lymphocyte subsets in the MS process is still obscure. This fact may be due to different methodological problems, such as the low number of cells, the small number of healthy controls studied and differences in technique. However, our limited comprehension of the true contribution of these studies to MS pathogenesis may be due to some unresolved underlying problems. First, it is important to keep in mind that MS is a more complicated process than EAE, so that the T cell role in MS may be more subtle and sophisticated than in the oversimplified animal model. In this sense, it may be useful to consider CSF lymphocyte studies as a part rather than the totality of the pathogenetic mechanisms of MS. In this case, knowledge of CSF T cell activation and the presence of a selective subset of T cells in CSF has greatly facilitated the comprehension of the immunological characteristics of the central nervous system. Secondly, it is likely that the core of the MS process takes place inside the white matter, and that what we can deduce from CSF studies may be just a reflection of the true biological damage, which is the basis for MS.

References

1. Lassman H, Wisniewky HM (1979) Chronic relapsing experimental allergic encephalomyelitis. Clinicopathological comparison with multiple sclerosis. Arch Neurol 36: 490-497

2. Manconi PE, Zaccheo D, Bugiani O et al (1976) T and B lymphocytes in cerebrospinal fluid. N Engl J Med 294:49
3. Manconi PE, Zaccheo D, Bugiani O et al (1978) Surface markers on lymphocytes from human cerebrospinal fluid. I. Predominance of T-lymphocytes bearing receptors for the Fc segment of IgG. Eur Neurol 17: 87-91
4. Allen JC, Sheremata W, Cosgrove JBR, Osterland K, Shea M (1976) Cerebrospinal fluid T and B kinetics related to exacerbation of multiple sclerosis. Neurology 26: 579-583
5. Goust JM, Chenais F, Carnes JE, Hames CG, Fudenberg HH, Hogan EL (1978) Abnormal T cell subpopulations and circulating immune complex in Guillain-Barré and multiple sclerosis. Neurology 28: 421-425
6. Huddleston JR, Oldstone MBA (1979) T suppressor (Tg) lymphocytes fluctuate in parallel with changes in the clinical course of patients with multiple sclerosis. J Immunol 123: 1615-1618
7. Manconi PE, Marrosu MG, Cianchetti C, Ennas MG, Mangoni A, Zaccheo D (1980) Lymphocyte subpopulations in cerebrospinal fluid and peripheral blood in multiple sclerosis. Acta Neurol Scand 62: 165-175
8. Sandberg-Wollheim M (1983) Lymphocyte populations in cerebrospinal fluid and peripheral blood of patients with multiple sclerosis and optic neuritis. Scand J Immunol 17: 575-581
9. Reiherz EL, Kung P, Goldstein G, Schlossman SF (1979) Separation of functional subsets of human T cells by a monoclonal antibody. Proc Natl Acad Sci USA 76: 4061-4065
10. Marrosu MG, Ennas MG, Murru MR, Marrosu G, Cianchetti C, Manconi PE (1983) Surface markers on lymphocytes from human cerebrospinal fluid. III. Identification by monoclonal antibodies. J Neuroimmunol 5: 325-331
11. Marrosu MG, Cianchetti C, Ennas MG (1986) Cerebrospinal fluid lymphocyte subpopulations in multiple sclerosis. Ital J Neurol Sci 7:101-105
12. Marrosu MG (1989) Cerebrospinal fluid T cell subsets in multiple sclerosis. In: Battaglia MA, Crimi G (eds) An Update on Multiple Sclerosis. Monduzzi Ed, pp 131-134
13. Marrosu MG, Cianchetti C, Mannu L et al (1989) Effect of corticosteroids on lymphocyte subpopulations in multiple sclerosis. In: Battaglia MA, Crimi G (eds) An Update on Multiple Sclerosis. Monduzzi Ed, pp 135-140
14. Comston A (1983) Lymphocyte subpopulations in patients with multiple sclerosis. J Neurol Neurosur Psychiatry 46: 105-114
15. Reunanen MI (1982) Spontaneous proliferation of cerebrospinal fluid mononuclear cells in multiple sclerosis. A longitudinal study. J Neuroimmunol 3: 275-283
16. Marrosu MG (1990) Imbalance of CSF lymphocyte subsets in multiple sclerosis. In: Marrosu MG, Cianchetti C, Tavolato B (eds) Trends in Neuroimmunology. Plenum Press, NY, pp 59-66
17. Cantrell DA, Smith KA (1984) The interleukin-2 T-cell system: a new cell growth model. Science 224: 1312-1316
18. Bellamy S, Calder VL, Feldmann M, Davison AN (1985) The distribution of interleukin-2 receptor bearing lymphocytes in multiple sclerosis: evidence for a key role of activated lymphocytes. Clin Exp Immunol 61: 248-256
19. Tournier-Lasserve E, Lyoncaen O, Roullet E, Bach MS (1987) Il -2 receptor and HLA class II antigens on cerebrospinal fluid cells of patients with multiple sclerosis and other neurological diseases. Clin Exp Immunol 67: 581-586
20. Serra HM, Krowka JF, Ledbetter JA, Pilarki LM (1988) Loss of CD45R (Lp 220) represents a post-thymic T cell differentiation event. J Immunol 140: 1435-1441
21. Konttinen YT, Bergroth V, Kinnunen E, Nordstrom D, Kouri T (1987) Activated T lymphocytes in patients with multiple sclerosis in clinical remission. J Neurol Sci 81: 133-139
22. Morimoto C, Letvin NL, Distase JA, Aldrich WR, Schlossman SF (1985) The isolation

and characterization of the human suppressor inducer T cell subset. J Immunol 134: 1508-1515
23. Morimoto C, Letvin NL, Boyd AW, Hagan M, Brown HM, Kornacki MN, Schlossman SF (1985) The isolation and characterization of the human helper inducer T cell subset. J Immunol 134: 3762-3769
24. Sanders ME, Makgoba MW, Sharrow SO, Stephany D, Springer TM, Young HA, Shaw S (1988) Human memory T lymphocytes express increased level of three cell adhesion molecules (LFA-3), CD2, and LFA-1) and three other molecules (UCHL 1, CDw29, and PgP-1) and have enhanced INF production. J. Immunol 140:1401-1407
25. Sanders ME, Makgoba MW, Shaw S (1988) Human naive and memory T cells: reinterpretation of helper-inducer and suppressor-inducer subsets. Immunol Today 9: 195-199
26. Chofflon M, Weiner HL, Morimoto C, Hafler DA (1989) Decrease of suppressor-inducer (CD4+2H4+) T cells in multiple sclerosis cerebrospinal fluid. Ann Neurol 25: 494-499
27. Salonen R, Ilonen J, Jagerroos H, Syrjala H, Nurmi T, Reunanen M (1989) Lymphocyte subsets in the cerebrospinal fluid in active multiple sclerosis. Ann Neurol 25: 500-502
28. Matsui M, Mori KJ, Saida T (1990) Cellular immunoregulatory mechanism in the central nervous system: characterization of noninflammatory and inflammatory cerebrospinal fluid lymphocytes. Ann Neurol 27: 647-651
29. Marrosu MG (1991) Decrease in memory (CDW29-high) cerebrospinal fluid cells in acute multiple sclerosis patients. Acta Neurol Scand 84: 487-490
30. Springer TA (1990) Adhesion receptors of the immune system. Nature 346: 425-434
31. Svenningsson A, Hansson GK, Andersen O, Andersson R, Patarroyo M, Stemme S (1993) Adhesion molecule expression on cerebrospinal fluid T lymphocytes: evidence for common recruitment mechanisms in multiple sclerosis, aseptic meningitis, and in normal controls. Ann Neurol 34: 155-161
32. Fleischer B, Kreth HW (1983) Clonal analysis of HLA-restricted virus-specific cytotoxic T lymphocytes from the cerebrospinal fluid in mumps memingitis. J Immunol 130: 2187-2190
33. Fleischer B, Marquardt P, Poser S, Kreth HW (1984) Phenotypic markers and functional characteristics of T lymphocyte clones from cerebrospinal fluid of patients with multiple sclerosis. J Neuroimmunol 7: 151-162
34. Santoli D, Defreitas EC, Sandberg-Wollheim M, Francis MK, Koprowki H (1984) Phenotypic and functional characterization of T cell clones derived from the cerebrospinal fluid of multiple sclerosis patients. J Immunol 132: 2386-2392
35. Clark RB, Dore-Duffy P, Donaldson JO, Pollard MK, Muirhead SP (1984) Generation of phenotypic helper/inducer and suppressor/cytotoxic T-cell lines from cerebrospinal fluid in multiple sclerosis. Cell Immunol 84: 409-414
36. Hafler DA, Buchsbaum M, Johnson D, Weiner HL (1985) Phenotypic and functional analysis of T cells cloned directly from the blood and cerebrospinal fluid of patients with multiple sclerosis. Ann Neurol 18: 451-458
37. Richert JR, McFarlin DE, Rose JW, McFarland HF, Greenstein JI (1983) Expansion of antigen-specific T cells from cerebrospinal fluid of patients with multiple sclerosis. J Neuroimmunol 5: 317-324
38. Weber WEJ, Buurman WA, Vandermeeren MMPP, Medaer RHJ, Raus JCM (1987) Fine analysis of cytolytic and natural killer T lymphocytes in the CSF in multiple sclerosis and other neurological diseases. Neurology 37: 419-425

Cerebrospinal fluid and serum soluble adhesion molecules in multiple sclerosis

M. Trojano, C. Avolio, I.L. Simone, M. Ruggieri and P. Livrea

Introduction

Multiple Sclerosis (MS) is a chronic demyelinating disease of the central nervous system (CNS) characterized by a multifocal inflammatory leukocyte infiltration and demyelination. It is widely accepted that an "activation" of CNS microvascular endothelial cells (EC) associated with blood-brain barrier (BBB) damage represents early event in demyelinating lesion development [1, 2]. However, the mechanisms responsible for the migration of lymphocytes through microvascular EC into the CNS have not been clearly identified. This traffic needs peripheral blood mononuclear cells (PBMC) activation, recruitment and strong adhesion to vascular ECs. Normally, vascular ECs have low adhesiveness for PBMC, but when stimulated by cytokines, such as interleukin-1, tumor necrosis factor and interferon gamma [3-5], released by inflammatory cells, they express surface adhesion proteins.

Recent efforts resulted in the identification, characterization and cloning of these cell-surface glycoproteins. The most important of these proteins are the intercellular adhesion molecule-1 (ICAM-1), the vascular cell adhesion molecule-1 (VCAM-1), and the endothelial leukocyte adhesion molecule-1 (ELAM-1 or E-selectin). It is likely that interaction between these EC adhesion proteins and activated T lymphocyte ligands or counter-receptors, such as leukocyte function antigen-1 (LFA-1), very late activation antigen 4 (VLA-4), and L- selectin (L-Se), may facilitate the capture, rolling, attachment, and extravasation of lymphocytes into inflamed tissues [6-9].

Recent studies showed that both the ICAM-1/LFA-1 and the VCAM-1/VLA-4 systems are involved in the pathogenesis of inflammatory demyelinating diseases, including experimental allergic encephalomyelitis (EAE), which is the animal model of MS. The expression of ICAM-1, VCAM-1 and E-selectin increases in mouse CNS microvascular ECs during the induction of EAE, and the upregulation of the adhesion molecules coincides with lymphocyte immigration into the CNS [10]. Some studies reported prevention or improvement of EAE by using specific mAbs for VLA-4 [11], for VLA-4 and/or its counterpart VCAM-1 [12], and for ICAM-1 [13]. However, mAbs against LFA-1 appeared to exacerbate EAE [14]. The major-

Institute of Clinical Neurology, University of Bari, Policlinico, Piazza Giulio Cesare, 70124 Bari, Italy

ity of CNS microvessels from MS patients express significant levels of ICAM-1, VCAM-1 and, to a lower degree, of E-selectin [15, 16].

Recently, soluble or circulating forms of these adhesion molecules have been detected in biological fluids such as serum and CSF by sensitive immunoassays. The origin of the soluble form of adhesion proteins is not entirely known; recent data suggest a major production by microvascular ECs and by mononuclear cells accompanied by a non-specific release from the damaged or inflamed tissue, or, alternatively, by a specific proteolytic cleavage of the membrane-bound parent molecule [17]. The putative sources of the intrathecally-produced soluble adhesion molecules could also include resident cells such as astrocytes [4] and cerebral perivascular cells [18]. The potential functions of soluble forms of adhesion molecules and their biological and diagnostic significance in MS have recently been the object of several investigations.

Intercellular adhesion molecule-1 (ICAM-1)

ICAM-1 (CD54) is a 90-Kd glycoprotein (Table 1) belonging to the Ig supergene family, with five extracellular Ig-like domains, a single transmembrane region and a short cytoplasmic tail. It is a ligand for at least 2 members of the CD18 family of leukocyte adhesion molecules: LFA-1 on lymphocytes, and MAC-1 on macrophages and neutrophils. ICAM-1 shows a very broad distribution; it is basically expressed on vascular endothelium and shows marked upregulation on most tissues in acute and chronic inflammatory diseases. ICAM-1 is expressed intensely by vascular endothelium during the worsening phase of RR EAE, less intensely during the relapses, and at diminished levels during remissions [8, 9]. It is also expressed by CNS capillaries at the edges of active plaques associated with MS [19]. The demonstration that anti-ICAM-1 mAbs can significantly decrease myelin basic protein (MBP)-induced lymphocyte proliferation, CNS leukocyte infiltration, and EAE clinical severity in rats [13], suggests a potential role of this molecule in the pathogenesis of EAE.

The soluble form of ICAM-1 contains most, if not all, of the five extracellular domains of the membrane-bound form, as well as the ability to bind specifically to LFA-1 [20]. Several groups reported that serum and CSF sICAM-1 levels differ in MS patients and in other inflammatory neurological diseases when compared with non inflammatory diseases. Some Authors (Tables 2 and 4) found significantly increased sICAM-1 levels in serum [21-23] and CSF [23, 24] of MS patients in clinically active phase of the disease, or showing CNS active lesions enhanced by Gd-DTPA on MRI [21], in comparison with MS in inactive phase and subjects with non-inflammatory neurological diseases. Other Authors, however, found that sICAM-1 serum levels in MS patients were not different from those in non-inflammatory neurological controls, and lower than those in viral and bacterial meningoencephalitis [25].

More recent longitudinal analyses of serum slCAM-1 levels also showed discordant results. Rieckmann et al. [26] found in a population of 29 RR MS patients, examined monthly over a period of 12 months, a cumulative sICAM-1 level in 18 ac-

Table 1. Membrane bound adhesion molecules

Molecule	Structure	Expression	Regulation	Ligands	Binds to	Putative role
ICAM-1 (CD54)	Immunoglob. superfamily, 5 Ig domains, 90-115 Kda	endothelial cells, B-cells, T-cells, monocytes, macrophages, microglia, astrocytes, dendritic cells	IL-1β, TNFα, IFN-γ, LPS	CD11a/CD18 (LFA-1), CD11b/CD18 (MAC-1 CR3), CD11c/CD18	lymphocytes, neutrophils, monocytes, NK-cells	leukocyte activation, adhesion, transmigration
VCAM-1 (CD106)	Immunoglob. superfamily, 6 Ig domains, 90-110 Kda	endothelial cells, macrophages, dendritic cells, bone marrow stromal cells, synovial cells	IL-4, IL-1β, TNFα	CD49d/CD29 (VLA-4 or α4β1 integrin), α4β7	lymphocytes, monocytes, eosinophils, basophils	leucocyte activation, adhesion, transmigration
E-selectin (CD62E)	selectin/ LECCAM family 107-115 Kda	endothelial cells	IL-1β, TNFα, LPS	sialylated Lewis x and a (sLe^x, sLe^a)	neutrophils, monocytes, subset of memory T-cells, eosinophils, basophils	leukocyte capture, rolling
L-selectin (CD62L)	selectin family, 75 Kda (on leucocytes), 90-100 Kda (on neutrophils)	lymphocytes, monocytes, neutrophils		phosphorylated mono and polysaccharides (PPME), sulfated polysaccharides (fucoidan), sLe^x, sLe^a	endothelial cells, lymph node high endothelial venules	leukocyte capture, rolling
P-selectin (CD61)	selectin family 140 Kda	platelets, endothelial cells	thrombin, histamin, terminal complement component, H_2O_2, endotoxin, TNFα	sulfated polysaccharides (fucoidan, heparin) sLe^x, sLe^a	neutrophils, monocytes	leukocyte adhesion, rolling

Table 2. Serum levels of soluble sICAM-1 in MS patients

References	Controls (ng/ml)	CP MS (ng/ml)	RR clinically inactive MS (ng/ml)	RR clinically active MS ng/ml)	RR MS with Gd-enhancing MRI lesions (ng/ml)	RR MS with Gd-non-enhancing MRI lesions (ng/ml)	Detect. limits (ng/ml)	Method
sICAM-1								
Sharief [23]	34.3% (> cut-off % Abs)	–	25% (> cut-off % Abs)	61% (> cut-off% Abs)	–	–	–	dot-blot
Tsukada [22]	173±56.3	–	187±33.4	299.2±71.1	–	–	0.625	ELISA
Jander [25]	305±140	–	–	270±70	–	–	0.3	ELISA
Hartung [21]	159.8±24.4	236.1±93.8	169.1±29.5	322.6±102.2	349.4±87.9	216.1±91.6	0.5	ELISA
Rieckmann [26]	218±68 (annual cumulative values)	–	225±82 (annual cumulatives values)	502±218 (annual cumulative values)	–	–	0.5	ELISA
Dore-Duffy [27]	146±41	360±181	196±81	198±116	–	–	8	ELISA
Trojano [in press]	252±68	–	–	284±60	272±43	317±86	0.47	ELISA

CP = chronic progressive; RR = relapsing remitting

tive MS patients significantly higher than in 11 stable patients; they also demonstrated that peak levels for sICAM-1 coincided with the onset of clinical relapses. Dore-Duffy et al. [27] showed high sICAM-1 serum levels in 14 CP MS patients, but not in 36 stable and in 8 active RR MS. Moreover, they found a decrease of these levels during the clinical exacerbations in 4 RR MS patients also followed longitudinally and monthly, over a period of 12 months.

Recently, we measured [28] sICAM-1 in the serum and CSF of 35 clinically active RR MS patients who underwent both lumbar puncture and Gd-DTPA-MRI within an interval of one week, and of 30 neurological controls of whom 16 had non-inflammatory diseases (NIND), 8 had bacterial meningitis (BM) and 5 AIDS dementia complex (ADC). We confirmed elevated serum sICAM-1 levels in clinically active RR MS patients and in inflammatory neurological controls compared to non-inflammatory controls; clinically active RR MS patients without Gd-DTPA enhancing lesions on MRI were characterized by significantly lower CSF/serum albumin (QAlbumin) values and mononuclear CSF cell content but also, unlike in a previous report [21], by higher serum sICAM-1 levels compared to patients with enhancing lesions. We found a strong correlation between QsICAM and QAlbumin in controls ($R=0.95$; $p<0.0001$) with a normal BBB function as well as in BM with a BBB damage ($R=0.82$; $p=0.001$) (Fig. 1), suggesting that CSF sICAM-1 levels are influenced by levels in serum and by BBB permeability properties. Unlike the correlations between QAlbumin and quotients of larger molecules like IgG (146 Kdalton), IgA (385 Kdalton) and IgM (970 Kdalton) [29], a linear approach for the correlation between QAlbumin (60 Kdalton) and QsICAM-1 (89 Kdalton) may be acceptable due to their similar size. Thus, the sICAM-1 index (CSF sICAM-1/serum sICAM-1:CSF Albumin/serum Albumin) [30] may be considered a reliable marker of intrathecal sICAM-1 production since it demonstrates sICAM-1 local production, and not only its passive diffusion through a defective BBB. By means of the sICAM-1 index, intrathecal sICAM-1 synthesis was demonstrated in MS and in the other demyelinating disease, ADC [25, 28].

A positive correlation was found between the number of CSF mononuclear cells and CSF sICAM-1 level [23, 24] or sICAM-1 index values [28], confirming that, at least partially, these cells are responsible for sICAM-1 local production. However, other CNS resident cells such as astrocytes [31] and microglia, other than blood-borne lymphocytes [18], seem to be involved in sICAM-1 intrathecal synthesis. Short term corticosteroid treatment [22, 28] led to a reduction in both the serum sICAM-1 levels and its intrathecal production.

Selectins

The selectin family includes three cell surface glycoproteins that are respectively found on endothelium (E-selectin; ELAM-1), leucocytes (L-selectin) and platelets/endothelium (P-selectin) (Table 1). Sequence data for each of the three molecules defined a type 1 transmembrane protein with a N-terminal lectin-like domain, and epidermal growth factor repeat, and a variable number of complement regulatory-like modules of about 60 amino acids each [6].

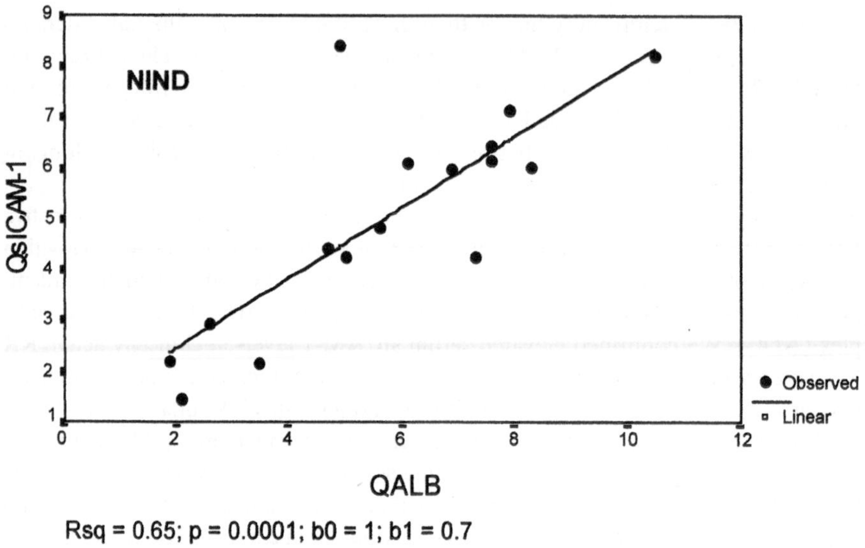

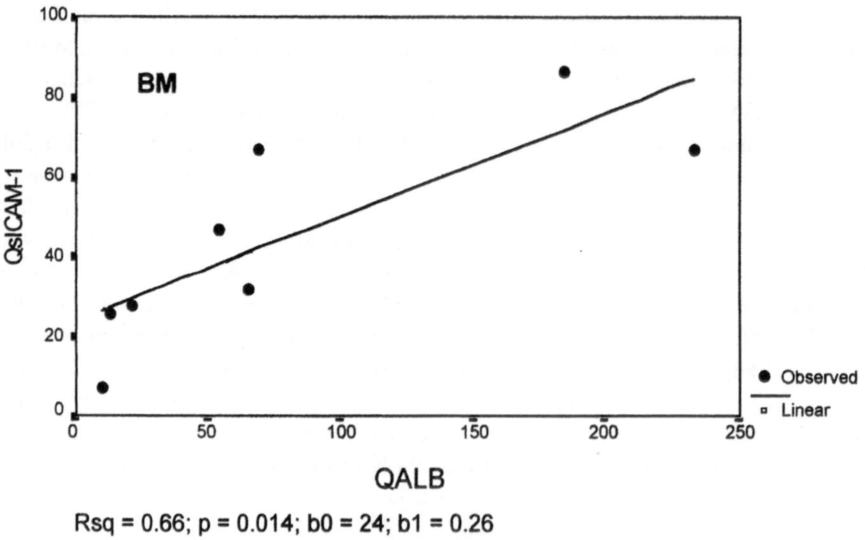

Fig. 1. QsICAM-1 vs. QAlb linear regression analysis plot in 16 patients with non inflammatory neurological diseases (NIND) and in 8 patients with bacterial meningitis (BM)

E-selectin is a 107-115 Kda glycoprotein expressed on endothelial cells (predominantly on post-capillary venules) after stimulation with cytokines IL-1β and TNFα, as well as bacterial endotoxin LPS. Studies *in vitro* have suggested that E-selectin binds mainly to neutrophils, but also to monocytes, a subset of memory T-cells, eosinophils and basophils [6].

P-selectin, approximately 140 Kda in size, (known also as platelet activation dependent granule-external protein – PADGEM – or granule membrane protein 140 - GPM140), is expressed on activated platelets. It has also been found in storage/secretory granules of endothelial cells; a variety of mediators including thrombin, istamin, terminal complement component, H202, endotoxin and TNFα stimulate its surface expression and at the same time induce an increased BBB permeability. P selectin binds different types of leukocytes including neutrophils and monocytes [6].

L-selectin has a size ranging from 75 Kda to 90-110 Kda. Unlike the other two selectins, L-selectin is constitutively expressed at the cell surface of most circulating human lymphocytes, neutrophils and monocytes; it facilitates the extravasion of leukocytes across lymph node high endothelial venules. All three molecules have been shown to bind a variety of natural as well as synthetic carbohydrates, which can be divided into three main categories: a) oligosaccharides related to syalilated Lewis x (sLe^x) and syalilated Lewis a (sLe^a); b) phosphorylated mono and polysaccharides; c) sulfated polysaccharides. Moreover, additional cell proteins may participate in the presentation of selectin ligands. Nevertheless molecular details of selectin-ligand-interaction have not yet been defined, but certainly the binding between different selectins and different ligands could involve different mechanisms. The selectin group of adhesion molecules seems to be of paramount importance at sites of inflammation, in the first stage of capture, rolling, and loose attachment of leukocytes to the luminal wall of vessels. Rolling leukocytes are arrested at the vascular lining under the influence of chemoattractants such as chemokines, C5a, and platelet-activating factor. Since microvessels from the CNS of MS patients express E-selectin, even if irregularly, at high levels its role in the lymphocyte migration seems to be important, especially during the initial inflammatory stage of demyelination. A recent study suggests that the anti-inflammatory effects of corticosteroids involve such a mechanism [32].

Although selectins are transmembrane glycoproteins, soluble forms have been detected in human sera from various inflammatory diseases [17]. The mechanisms of their production is unclear, but a proteolytic cleavage from the cell surface of the parent membrane bound molecules has been indicated as the most probable. Until now, soluble form of selectins has not been extensively studied in MS (Tables 3 and 4) as was the soluble form of ICAM-1. Rieckmann et al. [26] found high L-selectin serum levels in both active and stable MS compared with healthy donors, whereas there were no differences between groups for circulating ELAM-1 neither selectins were a associated with disease activity in a longitudinal evaluation. Hartung et al. [33] reported higher serum levels of sL-selectin in clinically active RRMS compared to inactive RR and CP patients and to healthy controls; by contrast, sELAM-1 levels did not differ. However, both selectins were higher in the sera of patients with MRI Gd-enhancing lesions compared to those wihout. Dore-Duffy et al. [27] found high serum and CSF ELAM-1 levels only in MS patients with CP course; CSF levels did not correlate with serum levels; when patients experienced a clinically documented exacerbation serum sELAM-1 level appeared to go down. Finally, Tsukada et al. [34] detected higher levels of sE-selectin in serum and CSF of active RR and

Table 3. Serum levels of soluble E-Selectin, L-Selectin and VCAM-1 in MS patients

References	Controls (ng/ml)	CP MS (ng/ml)	RR clinically inactive MS (ng/ml)	RR clinically active MS (ng/ml)	RR MS with Gd-enhancing MRI lesions (ng/ml)	RR MS with Gd-non-enhancing MRI lesions (ng/ml)	Detect. limits (ng/ml)	Method
sE-SELECTIN								
Rieckmann [26]	28.9±8.8 (annual cumulative value)	–	30.2±12.8 (annual cumulative value)	34.8±11.2 (annual cumulative value)	–	–	1.6	ELISA
Dore-Duffy [27]	7.4±8	15.1±13	8±4	5.8±4	–	–	0.1	ELISA
Tsukada [34]	30.5±12.4	54.7±19.1	44.7±12.7	75.2±12.1	–	–	0.1	ELISA
Hartung [33]	21.9±8.1	21 (median)	19 (median)	21 (median)	24±9.9	20.9±9.8	1.6	ELISA
sL-SELECTIN								
Rieckmann [26]	1312±145 (annual cumulative value)	–	1889±508 (annual cumulative value)	2002±412 (annual cumulative value)	–	–	0.3	ELISA
Hartung [33]	793±184	769 (median)	768 (median)	995 (median)	1130±272	793±207	0.3	ELISA
sVCAM-1								
Dore-Duffy [27]	428±84	400±158	399±97	412±158	–	–	1.3	ELISA
Matsuda [37]	402.5±161.5	1254±885.3	387.6±118.6	691.8±288.6	–	–	140	ELISA
Hartung [33]	593.4±125	686 (median)	598 (median)	877 (median)	1014±275	634±191	1.8	ELISA

CP = chronic progressive; RR = relapsing - remitting

Table 4. CSF levels of soluble ICAM-1, E-Selectin and VCAM-1 in MS patients

References	Controls (ng/ml)	CP MS (ng/ml)	RR clinically inactive MS (ng/ml)	RR clinically active MS (ng/ml)	RR MS with Gd-enhancing MRI lesions (ng/ml)	RR MS with Gd-non-enhancing MRI lesions (ng/ml)	Detect. limits (ng/ml)	Method
sICAM-1								
Tsukada [24]	3.6±1.1	8.1±2.6	3.2±1.3	10.2±1.5	–	–	0.625	ELISA
Jander [25]	0.5 - 2 (0.75±0.15 sICAM-1 Index)	–	–	0.5 - 2 (1±0.4 sICAM-1 Index)	–	–	0.3	ELISA
Sharief [23]	n.d.	–	20% of pts.	53% of pts.	–	–	–	dot-blot
Dore-Duffy [27]	n.d.	n.d.	n.d.	n.d.	–	–	8	ELISA
Trojano (in press)	1.4±0.78 (0.9±0.3 sICAM-1 Index)	/	/	1.75±0.87 (1.3±0.5 sICAM-1 Index)	1.77±0.99 (1.2±0.4 sICAM-1 Index)	1.71±0.5 (1.5±0.7 sICAM-1 Index)	0.47	ELISA
sE-SELECTIN								
Dore-Duffy [27]	0.04±0.01	0.08±0.04	0.06±0.01	0.06±0	–	–	0.1	ELISA
Tsukada [34]	< 0.1	< 0.6	< 0.2 (2/15)	< 1 (8/24)	–	–	0.1	ELISA
sVCAM-1								
Dore-Duffy [27]	1.1±0.7	8±3	4±1	4±2	–	–	1.3	ELISA
Matsuda [37]	4.8±1.0	20.9±9.3	5.4±2.3	14.3±4.5	–	–	2.8	ELISA

n.d. = not detectable; sICAM Index = sICAM-1 CSF/sICAM-1 serum: Albumin CSF/Albumin serum
CP = chronic progressive; RR = relapsing - remitting

CP MS compared with patients in remission and with non-inflammatory controls. Corticosteroid treatment reduced both serum and CSF ELAM-1 levels.

Vascular cell adhesion molecule-1 (VCAM-1)

The evidence of the presence of a third cytokine-inducible endothelial adhesion molecule, immunochemically distinct from previously described structures, has been clearly demonstrated over the past few years. This new molecule was designated as vascular cell adhesion molecule-1 (VCAM-1), and included as a member of the immunoglobulin superfamily [6].

VCAM-1 contains six-seven Ig domains and it is 90-110 Kda in size; the N-terminal domain (domain 1) and the fourth domain are similar in amino acid sequence, and appear to function in leukocyte adhesion. Endothelial cells, after stimulation with IL-1β, TNFα and IL-4, express VCAM-1. VCAM-1 is also found on several other non-vascular cell types, including dendritic cells in lymph node and skin, bone marrow stromal cells and synovial cells in inflamed joints.

The selective ligand for the VCAM-1 is represented by a β1 integrin, better known as α4β1 (α-chain 150 Kda and β-chain 120-130 Kda in size), or very late activation-4 (VLA-4;CD49d/CD29) molecule, expressed on resting lymphocytes, monocytes and neural crest derived cells. The integrin family includes three main subfamilies distinguished according to the presence of subunit β1 (CD29), β2 (CD18) and β3 (CD61); β1 integrins (VLA-4) are involved in the binding of cells to extracellular matrix, β2 integrins (LFA-1) are involved in leukocyte adhesion to endothelium or to other immune cells, β3 integrins are involved in the interactions of platelets and neutrophils at inflammatory sites or sites of vascular damage. All of them are transmembrane receptors able to transduce signals from the outside to the inside of the cell but also, as described for the LFA-1, to transduce signals from the cytosol to generate changes in extracellular functions, such as adhesion [35]. Endothelial VCAM-1 has been demonstrated to support the adhesion of lymphocytes, monocytes, eosinophils and basophils, but not neutrophils.

In EAE of mice, VCAM-1 as L-selectin and ICAM-1 is upregulated during the induction phase of the disease; however, compared to ICAM-1, VCAM-1 expression was found on a lower percentage of CNS vessels, with delayed kinetics. One explanation is that VCAM-1 can be induced by cytokines (IL-4, and less potently by IL-1β and TNFα) different from those (IL1β, TNFα, and less potently by IFN-γ) which induce ICAM-1. Thus, the delayed expression of VCAM-1 could have been the result of a shift from Th1 to Th2 lymphocytes [10]. The interaction of ICAM-1 and VCAM-1 with their respective ligands promotes strong adhesion of leukocytes to endothelium. Moreover, ICAM-1, but not VCAM-1, seems to play a role in their transendothelial migration [36]. ICAM-1 and its ligand LFA-1 play a major role in the adhesion of LFA-1-positive cells to resting ECs, whereas VCAM-1/VLA-4 appears to be more important in binding to actived ECs [15].

Yednock et al. [11] showed that, *in vitro*, mAbs for VLA-4 could block the binding of lymphocytes and monocytes to inflamed EAE in brain vessels, and *in vivo* anti-VLA-4 could effectively prevent the accumulation of leukocytes into CNS, and thereby the development of EAE. Tanaka's group [12] showed that specific mAbs to either VLA-4 or VCAM-1 partially inhibited the binding of MBP-specific T-cell clone to brain endothelial cells of SJL mice *in vitro*. Nevertheless, the binding was not inhibited if previously increased by incubation with IL-1β and TNFα. This could mean that adhesion molecules other than VLA-4_a/VCAM-1 are involved in the binding between T-cell clone to activated brain endothelium. MS microvessels, above all in periplaque zones, express significant levels of VCAM-1 together with ICAM-1, MHC class II antigens and urokinase plasminogen activator receptor [15].

A soluble form of the VCAM-1 molecule has been recently quantified in serum and CSF from MS patients (Tables 3 and 4). Dore-Duffy et al. [27] reported higher sVCAM-1 levels in the CSF of MS patients compared to controls; these levels were higher in CPMS than in RR MS; no significant differences were found in serum levels between MS and control groups. Matsuda [37] demonstated that CSF and serum sVCAM-1 levels were significantly elevated in active RR as well as in CP MS compared to inactive RR patients and healthy individuals. Hartung et al. [33] found higher sVCAM-1 in sera of MS patients with clinically active disease of Gr enhancing-MRI lesions compared to those with stable MS and with progressive course.

Conclusion

The biological function of soluble adhesion molecules in MS is not completely known. However, the previous reported results suggest that an active immune reaction involving the production of adhesion molecules occurs in serum and CSF of MS patients, especially during an exacerbation and the chronic progression of the disease. It is possible that the release in serum of these cell surface molecules could be involved in fundamental regulatory mechanisms to control adhesive interactions between Ec and the immunocompetent cells, and therefore focal accumulation of activated leukocytes into CNS [17, 38].

Circulating serum adhesion molecules, by competing with membrane-bound forms for the respective leukocyte ligands, down-regulating cytokine-induced increased expression, and thus facilitating detachment or preventing attachment could either increase transendothelial migration of leukocytes into CNS or, on the contrary, restrict leukocyte accumulation at the inflammed BBB. Moreover, the binding of soluble molecules to their target cells could have a co-stimulatory effect and a transmembrane signalling role in lymphocyte-endothelium and act as activating stimuli. The recombinant soluble E-selectin was likewise demonstrated to be a neutrophil chemoattractant [6]. The data *in vivo* until now reported, together with the results of a recent *in vitro* study on cultured cerebral endothelium [39], seem to indicate that the major role of serum adhesion molecules could be to block cell adhesion at the inflammatory or activated endothelium, and to reduce the cell migration into CNS during acute inflammatory events in MS.

Until now, only a few studies, except for sICAM-1, have been carried out in order to investigate the CSF levels of the adhesion molecules during the MS course.

Therefore, the potential functions and biological role of circulating CSF adhesion molecules can at present only be the subject of speculation. It was hypothesized that the intrathecally produced sICAM-1 could allow the initiation of a local immune response by facilitating contact between T cells and antigen-presenting cells, through binding to its ligand LFA-1, and also providing a co-stimulatory signal for T cell activation [40].

In conclusion, it appears that the monitoring of soluble adhesion molecules levels in serum and CSF, since it provides important information on endothelial inflammatory states, might be used as potential diagnostic and prognostic test in diseases which have associated vascular endothelial pathology, such as MS. Moreover, soluble molecules could, probably, represent a reliable parameter for the monitoring of the different disease phases. The apparently divergent results obtained in the various studies may be due to differences in methodology and time of sampling in relation to disease activity and, therefore, warrant further longitudinal evaluations supported by clinical and Gd-MRI investigations. Further understanding of the regulatory role played by soluble adhesion proteins in EC activation may provide additional insights on the pathophysiology of MS and the development of new therapeutic approaches to MS.

References

1. Shimuzu Y, Newman W, Tanaka Y et al (1992) Lymphocyte interaction with endothelial cells. Immunol Today 13:106-112
2. Grossman RI, Braffman BH, Brorson JR et al (1988) Multiple Sclerosis: serial study of gadolinium-enhanced MR imaging. Radiology 169:117-122
3. Pober JS, Gimbrone MA, Lapierre LA et al (1986) Overlapping patterns of activation of human endothelial cells by interleukin-1, tumor necrosis factor and immune interferon. J Immunol 137:1893-1896
4. Fabry Z, Waldschmidt MN, Hendrickson D et al (1992) Adhesion molecules on murine brain microvascular endothelial cells: expression and regulation of ICAM-1 and Lgp 55. J Neuroimmunol 36:1-11
5. Wong D, Dorovini-Zis K (1992) Upregulation of intercellular adhesion molecule-1 (ICAM-1) expression in primary cultures of human brain microvessel endothelial cells by cytokines and lipopolysaccharides. J Neuroimmunol 39:11-22
6. Bevilacqua MP (1993) Endothelial-leukocyte adhesion molecule. Annu Rev Immunol 11:767-804
7. Springer TA (1994) Traffic signals for lymphocyte recirculation and leukocyte emigration: the multiple step paradigme. Cell 76:301-314
8. Wilcox CE, Ward AMV, Evans A, Baker D, Rothlein R, Turk JL (1990) Endothelial cell expression of the intercellular adhesion molecule-1 (ICAM-1) in the central nervous system of guinea pigs during acute and chronic relapsing experimental allergic encephalomyelitis. J Neuroimmunol 30:43-51
9. Cannella B, Cross AH, Raine CS (1991) Adhesion-related molecules in the central nervous system. Upregulation correlates with inflammatory cell influx during relapsing experimental autoimmune encephalomyelitis. Lab Invest 1:23-31
10. Dopp JM, Breneman SM, Olschowka JA (1994) Expression of ICAM-1, VCAM-1, L-selectin, and leukosialin in the mouse central nervous system during the induction and re-

mission stages of experimental allergic encephalomyelitis. J Neuroimmunol 54:129-144
11. Yednock TA, Cannon C, Fritz LC, Sanchez-Madrid F, Steinman L, Karin N (1992) Prevention of experimental autoimmune encephalomyelitis by antibodies against $\alpha 4\beta 1$ integrin. Nature 356:63-66
12. Tanaka M, Stato A, Makino M, Tabira T (1993) Binding of an SJL T cell clone specific for myelin basic protein to SJL brain microvessel endothelial cells is inhibited by anti-VLA-4 or its ligand, anti-vascular cell adhesion molecule 1 antibody. J Neuroimmunol 46:253-258
13. Archelos JJ, Jung S, Maurer M et al (1993) Inhibition of experimental autoimmune encephalomyelitis by an antibody to the intercellular adhesion molecule ICAM-1. Ann Neurol 34:145-154
14. Cannella B, Cross AH, Raine CS (1993) Anti-adhesion molecule therapy in experimental autoimmune encephalomyelitis. J Neuroimmunol 46:43-56
15. Washington R, Burton J, Todd III RF, Newman W, Dragovic L, Dore-Duffy P (1994) Expression of immunologically relevant endothelial cell activation antigens on isolated central nervous system microvessels from patients with multiple sclerosis. Ann Neurol 35:89-97
16. Raine CS, Lee SC, Scheinberg LC, Duijvestijn AM, Cross AH (1990) Adhesion molecule on endothelial cells in the central nervous system: an emerging area in immunology of multiple sclerosis. Clin Immunol Immunopathol 57:173-187
17. Pigott R, Dillon LP, Hemingway IH, Gearing AJH (1992) Soluble forms of E-selectin, ICAM-1 and VCAM-1 are present in the supernatants of cytokine activated cultured endothelial cells. Biochem Biophys Res Commun 187/2:584-589
18. Cannella B, Cross AH, Raine CS (1990) Upregulation and coexpression of adhesion molecules correlate with relapsing autoimmune demyelination in the central nervous system. J Exp Med 172:162-170
19. Sobel RA, Mitchell ME, Fondren G (1990) Intercellular adhesion molecule-1 (ICAM-1) in cellular immune reactions in the human central nervous system. Am J Pathol 136 (6): 1309-1316
20. Rothlein R, Mainolfi EA, Czajkowski M, Marlin DS (1991) A form of circulating ICAM-1 in human serum. J Immunol 147 (11):3788-3793
21. Hartung HP, Michels M, Reiners K, Seeldrayers P, Archelos JJ, Toyka KV (1993) Soluble ICAM-1 serum levels in multiple sclerosis and viral encephalitis. Neurology 43:2331-2335
22. Tsukada N, Miyagi K, Matsuda M, Yanagisawa N (1993) Increased levels of circulating intercellular adhesion molecule-1 in multiple sclerosis and human T-lymphotropic virus 1-associated myelopathy. Ann Neurol 33:646-649
23. Sharief MK, Noori MA, Ciardi M, Cirelli A, Thompson EJ (1993) Increased levels of circulating ICAM-1 in serum and cerebrospinal fluid of patients with active multiple sclerosis. Correlations with TNF-alpha and blood-brain barrier damage. J Neuroimmunol 43:15-21
24. Tsukada N, Matsuda M, Miyagi K, Yanagisawa N (1993) Increased levels of intercellular adhesion molecule-1 (ICAM-1) and tumor necrosis factor receptor in the cerebrospinal fluid of patients with multiple sclerosis. Neurology 43:2679-2682
25. Jander S, Heidenreich F, Stoll G (1993) Serum and CSF levels of soluble intercellular adhesion molecule-1 (ICAM-1) in inflammatory neurologic diseases. Neurology 43:1809-1813
26. Rieckmann P, Martin S, Weichselbraun I et al (1994) Serial analysis of circulating adhesion molecules and TNF receptor in serum from patients with multiple sclerosis: cICAM-1 is an indicator for relapse. Neurology 44:2367-2372
27. Dore-Duffy P, Newman W, Balbanov R et al (1995) Circulating, soluble adhesion pro-

teins in cerebrospinal fluid and serum of patients with multiple sclerosis: correlation with clinical activity. Ann Neurol 37:55-62

28. Trojano M, Avolio C, Simone IL et al. Soluble Intercellular Adhesion Molecule-1 in serum and cerebrospinal fluid (CSF) of clinically active relapsing-remitting multiple sclerosis: correlation with Gd-DTPA Magnetic Resonance Imaging-enhancement and CSF findings. Neurology (in press)
29. Reiber H, Felgenhauer K (1987) Protein transfer at the blood cerebrospinal fluid barrier and the quantitation of the humoral immune response within the central nervous system. Clin Chim Acta 163:319-328
30. Link H, Tibbling G (1977) Principles of albumin in IgG analysis in neurological disorders. III. Evaluation of IgG synthesis within the central nervous system in Multiple Sclerosis. Scand J Clin Lab Invest 37:397-401
31. Vollmer T, Dalton J, Klotz K, Eddstein J (1992) Astrocyte-lymphocyte binding: the role of immune cytokines,and ICAM-1. Neurology 42 (Suppl. 3): 158-159
32. Bevilacqua MP, Nelson RM (1993b) Selectins. J Clin Invest 91:379-387
33. Hartung PH, Reiners K, Archelos JJ et al (1995) Circulating adhesion molecules and Tumor Necrosis Factor receptor in multiple sclerosis: correlation with Magnetic Resonance Imaging. Ann Neurol 38:186-193
34. Tsukada N, Miyagi K, Matsuda M, Yanagisawa N (1995) Soluble E-selectin in the serum and cerebrospinal fluid of patients with multiple sclerosis and human T-lymphotropic virus type 1-associated myelopathy. Neurology 45:1914-1918
35. Springer TA (1990) Adhesion receptors of the immune system. Nature 346:425-433
36. Oppenheimer-Marks N, Davis LS, Tompkins Bogue D et al (1991) Differential utilization of ICAM-1 and VCAM-1 during the adhesion and transendothelial migration of human T lymphocytes. J Immunol 147:2913-2921
37. Matsuda M, Tsukada N, Miyagi K, Yanagisawa N (1995) Increased levels of soluble vascular cell adhesion molecule-1 (VCAM-1) in the cerebrospinal fluid and sera of patients with multiple sclerosis and human T lymphotropic virus type-1-associated myelopathy. J Neuroimmunol 59:35-40
38. Gearing AJH, Hemingway I, Pigott R, Hughes J, Rees AJ, Cashman SJ (1992) Soluble forms of vascular adhesion molecule, E-selectin, ICAM-1, and VCAM-1: pathological significance. Ann NY Acad Sci 667:324-331
39. Rieckmann P, Michel U, Albrecht M, Bruck W, Wockel L, Felgenhauer K (1995) Soluble forms of intercellular adhesion molecule-1 (ICAM-1) block lymphocyte attachment to cerebral endothalial cells. J Neuroimmunol 60:9-15
40. Van Seventer GA, Shimizu Y, Horgan KJ, Shaw S (1990) The LFA-1 ligand ICAM-1 provides an important costimulatory signal for T cell receptor-mediated activation of resting T cells. J Immunol 144:4579-4586

Cytokines in multiple sclerosis cerebrospinal fluid and serum

P. GALLO AND B. TAVOLATO

Introduction

Under normal conditions, cytokines are not readily detected in biological fluids, such as serum and cerebrospinal fluid (CSF), because they have a very short lifetime, act mainly as autocrine/paracrine growth/immunomodulating factors, and are absorbed instantly at the tissue/cell target level by high affinity receptors (the biological features of cytokines are summarized in Table 1).

Table 1. Biological features of cytokines

- Polypeptides or glycoproteins with a low molecular weight (8-30 kD)
- Active at picomolar concentrations
- Bind to highly specific cellular receptors having high affinity and expressed at low concentrations on target cells
- Constitutive production of cytokines is usually low or absent; their production is regulated by various/simultaneous inducing/suppressive stimuli at the transcription or translation level
- Cytokine production is transient and radius of action is short (the typical action is autocrine and paracrine); however, when produced in large amounts in tissue or by circulating cells, they may exert hormone-like activities and can be detected in the biological fluids
- Cytokine effects are pleiotropic (broad spectrum of target cells) and can be understood only within the cytokine network
- Many important immunomodulatory/regulatory functions are shared by several cytokines

However, when cytokines are produced in large amounts in tissues or by circulating cells, they may be detected in body fluids, thus suggesting hormone-like activities. Indeed, many studies over the last ten years have found levels of cytokines in the serum of patients with systemic infectious, inflammatory and neoplastic diseases (Table 2). In the central nervous system (CNS), pro-inflammatory cytokines and growth factors are produced by both invading/infiltrating immune system cells (i.e. lymphocytes, monocytes/macrophages, granulocytes), and activated/reactive glial cells (i.e. astrocytes, microglial cells) (see Table 3).

Several studies have shown that cytokines accumulate in the CSF, especially in

Institute of Neurology, II Neurologic Clinic, University of Padua, Via E. Vendramini 7, 35137 Padua, Italy

Table 2. Pathological conditions in which the presence of one or more cytokines has been demonstrate in the serum of patients

Condition	IL-1β	TNFα	IL-6	sIL-2R
Rheumatoid arthritis	+	+	+	+
Parasitic infections		+		
Bacterial infections/Sepsis	+	+	+	
Severe burns	+		+	
HIV-1 infection		+		+
HTLV-I infection				+
Lymphoid malignancies			+	
Hemodialysis	+			+
Transplant rejections				+

Table 3. Immunological functions of astrocytes and microglial cells

	Astrocytes	**Microglia**
Phagocytosis	–	+
MHC-II[a] expression		
basal	–	+
IFNγ-induced	+	+
MHC-I[a] expression	–	+
Antigen processing	+	+
Antigen presentation	+	+
Macrophage phenotype	–	+
B7 expression	–	+
Cytokines		
• IL-1β	+	+
• TNFα	+	+
• IL-6	+	+
• M-CSF	+	+
• G-CSF	+	–
• GM-CSF	+	–
• IL-10	–	+
• IL-13	+	+
Growth Factors		
• NGF	+	?
• FGFb	+	+
• PDGF	+	?
• TGF	+	?

inflammatory and infectious CNS diseases, and the probability of detecting them increases with the acuteness and severity of the disease. Cytokines not only modulate immune system cell functions, but also may play a major role in stimulating glial cell proliferation and activation, and in inducing cytotoxic effector mechanisms, which may have a pathological impact on brain tissue, such as nitric oxide and quinolinic acid production, and proteinase activation.

At present, more than 100 pubblications describe cytokine detection in serum and CSF of patients with nervous system inflammatory and infectious diseases: about 50% of these reports involve multiple sclerosis (MS) patients. The basic objectives of these studies were to: 1) find a marker of disease activity and/or progression that could be used for diagnostic and/or prognostic purposes; 2) throw some light on the immune-mediated mechanism(s) responsible for peripheral T cell activation, loss of tolerance to CNS antigens, and white matter damage (demyelination).

Here, we review the data of the literature on the most investigated cytokines in MS (i.e. IL-1β, IL-2, sIL-2R, IL-6 and TNFα), and deduce that, although a general agreement exists that both IL-2 and sIL-2R are frequently increased in MS serum, no useful information can be taken for clinical, diagnostic and prognostic purposes.

Interleukin-1 beta (IL-1ß)

IL-1β may be produced intracerebrally by cells of the monocyte/macrophage lineage (MØ), including microglial cells, and by astrocytes. As IL-1-receptor expression has been well documented in brain tissue, the possibility that IL-1β might play a role in reactive gliosis was also suggested.

However, only one of the six major studies on IL-1β (Table 4) described frequent detection of this cytokine in MS CSF [1]; using a radioimmunoassay (RIA) with a detection limit of 50 pg/ml, Hauser et al. [1] found increased IL-1β levels not only in MS, but also in about one third of the CSF from patients with inflammatory (IND) and non-inflammatory (NIND) neurological diseases.

The frequency of positive MS CSF was much higher in active disease (85% acute relapsing, 50% chronic-progressive) than in stable disease (20%). The other

Table 4. IL-1β

Authors	Ref.	Methods	d.l.	MS CSF pos/total	OND pos/total
Gallo	[6]	Bioassay	0.1 U/ml	0/30 RR	vir. mening. 10/15
Hauser	[1]	RIA	50 pg/ml	28/33 acuteRR	IND 10/30
				11/20 active CP	NIND 28/102
				2/10 stable	GBS 0/6
Maimone	[2]	ELISA, Cistron Biotech.	40 pg/ml	2/34 (26 active)	IND 1/17
					NIND 0/35
Peter	[3]	ELISA, Cistron Biotech.	20 pg/ml	0/50 CP	NIND 0/19
Tsukada	[4]	ELISA, Otsuka Biotech.	?	0/20 acute RR	GBS 0/8
				0/9 active CP	CIDP 0/7
					NIND 0/11
Weller	[5]	ELISA, Cistron Biotech.	20 pg/ml	0/9 RR	GBS 0/9
				0/11 CP	CIDP 0/8
					vir. enceph. 0/7

five investigations, four of which employed commercial ELISA kits [2-5] and one a bioassay [6], obtained completely negative results (a total number of 175 MS patients with active disease, 70 chronic-progressive, and 105 relapsing-remitting, were studied). IL-1β could be detected only in patients with bacterial or viral meningitis but not in patients with Guillain-Barré syndrome (GBS), chronic inflammatory demyelinating polyneuropathy (CIDP), or NIND.

It is interesting to note that both the ELISA and the bioassay were reported to have a higher sensitivity (20-40 pg/ml) than RIA (50 pg/ml). The above data taken together cannot be analysed statistically, nor do they indicate any clinical advantage of IL-1 measurement in MS patients.

Interleukin-6 (IL-6)

Paired CSF and serum specimens from five large MS patient series were found to be completely negative for the presence of IL-6 [7-11]; however, four studies using four different detection methods (RIA, 7DT1 hybridoma, B9 hybridoma, ELISA) [1, 2, 5, 12], reported various degrees of positivity, ranging from 40% [1] to 100% [5] (Table 5). These discrepancies could not be explained by sensitivity differences in

Table 5. IL-6

Authors	Ref.	Methods	d.l. pos/total	MS CSF pos/total	OND CSF
Frei et al.	[7]	7TD1 hybridoma			
Houssiau	[8]	7TD1 hybridoma	1.0 U/ml (5pg/ml)	1/30	vir. mening. 25/42 bact. mening. 10/18 HSV enceph. 9/24
Leppert	[9]	7TD1 hybridoma	0.15 U/ml	4/42 RR	brain tumors 21/24 vir. mening. 19/22
Gallo	[10]	7TD1 hybridoma	0.15 U/ml	2/40 RR	vir. mening. 12/15 HSV-1 enceph 5/5
Hauser	[1]	RIA	20 pg/ml	18/45 active 1/15 inactive	OND (?) 23/25
Frei	[12]	7TD1 hybridoma	0.1 U/ml	13/27	vir. mening. 17/17 headache 3/15
Maimone	[2]	B9 hybridoma	1.0 pg/ml	5/10 relapsing 4/16 CP 1/8 stable	IND 9/19 NIND 3/43
Weller	[5]	ELISA British Biotech.	30 pg/ml	9/9 RR 3/11 CP	GBS 7/9 HIV enceph. 13/13 vir. mening. 7/7 bact. mening. 10/10 carc. mening. 27/30
Paemen	[11]	7TD1 hybridoma	3 pg/ml	0/55	optic neur. 0/46 IND 3/27 NIND 1/62

the techniques used. For instance, using a bioassay based on the proliferation of the IL-6-dependent 7DT1 hybridoma with a detection limit of 3 pg/ml, Paemen et al. [11] obtained unequivocal negative results, while Weller et al. [5] found 100% positivity with a commercial ELISA kit, having a detection limit of 30 pg/ml. Interestingly, the discrepancy observed among the 8 studies taken into consideration seems to be a peculiarity of MS patients, because comparable IL-6 levels were observed in the CSF of IND patients (especially viral and bacterial meningo-encephalitis), and IL-6 was only rarely found in patients with NIND. Hauser et al. [1] detected IL-6 in 23 of 35 (65%) patients with "other neurological disease", but provided no clinical information on these subjects. According to Frei et al. [12], it is possible to detect IL-6 in serum after warming up the samples, a condition, however, that cannot be considered physiological.

Tumor necrosis factor alpha (TNFα)

Interest in TNFα was kindled by experimental studies that described: 1) its potential demyelinating activity *in vitro* at high concentrations (i.e. 1,000- 10,000 U/ml); 2) its production by activated astrocytes and microglial cells; 3) the presence of TNFα-positive astrocytes in active demyelinating lesions. A role for this cytokine in myelin breakdown *in vivo* was advanced and, consequently, its intrathecal synthesis was extensively investigated not only in MS, but also in other conditions of inflammatory demyelination of the peripheral and central nervous systems, such as HIV-related encephalitis/leukoencephalopathy, GBS and CIDP.

At least five studies [3, 5, 13-15] failed to detect TNFα in both serum and CSF from patients with active relapsing-remitting (RR) or chronic-progressive (CP) MS, using either ELISA or bioassay; three of these were conducted in a large number of patients [3, 14, 15]. On the other hand, two investigations [1, 4] found increased levels of this cytokine in almost all the MS CSF tested. Despite its high detection limit (50 pg/ml), the RIA method used by Hauser et al. [1] showed the highest positivity (i.e. 90% of patients with active disease, and 80% of patients with stable disease had detectable TNFα level in their CSF). The discrepancies observed in the other groups of patients (Table 6) make the analysis and interpretation of TNFα even more difficult. For instance, in the study by Hauser et al. [1], 30% of the patients with NIND, 5/5 patients with stroke, 2/4 patients with migraine, and 5/13 patients with amyotrophic lateral sclerosis (ALS) had detectable TNFα levels in the CSF. Sharief et al. [16, 17] found TNFα in 50% of the active CP MS, but in 0/8 of the ALS patients. Maimone et al. [2] detected this cytokine in 7/16 CP (but in only 1/10 relapsing and 0/8 remitting MS), 30% of IND and 7% of NIND patients (but no clinical information was given about these patients).

Some of theses studies included GBS patients, in whom the percentage of TNFα-positive CSF varied from 0% [5] to 99% [4]; in the study by Franciotta et al. [15] 2 out of 2 patients were positive. A similar striking discrepancy is observed when the data thus far obtained in HIV-1-infected patients are examined (no TNFα positive CSF was found by Gallo et al. [14] and Weller et al. [5], while Franciotta et al. [15] re-

Table 6. TNF-α

Authors	Ref.	Methods	d.l.	MS CSF pos/total	OND CSF pos/total
Leist	[13]	L-M cell bioassay	≅0.3 U/ml	0/7	bact. mening. 3/7 vir. mening. 0/7
Gallo	[14]	L-M cell bioassay ELISA, Biokine TCS	≅0.5 U/ml 10 pg/ml	0/45 RR 0/15 CP	HIV infection 0/50 vir mening. 0/5
Franciotta	[15]	ELISA, Biokine TCS	10 pg/ml	0/27 RR 3/23 CP	HIV infection 5/5 GBS 2/2
Hauser	[1]	RIA	50 pg/ml	12/13 acute RR 6/8 active CP 7/9 stable CP	ALS 5/13 migraine 2/4 IND 11/19 NIND 12/18 stroke 5/5
Maimone	[2]	ELISA Cistron Biotech.	40 pg/ml	7/16 CP 1/18 RR	IND (?) 5/17 NIND (?) 3/41
Peter	[3]	ELISA, Biokine TCS	15 pg/ml	0/50 CP	dementia 0/15
Sharief	[16]	ELISA, Sandwich	2 U/ml	17/32 CP 0/20 stable	bact. mening. 3/6 GBS 1/7 ALS 0/8 sarcoidosis 0/5
Tsukada	[4]	ELISA, home-made	3.4 pg/ml	21/22 RR 8/9 CP	GBS 7/8 CIDP 5/7
Weller	[5]	ELISA, British Biotech.	40 pg/ml	0/9 RR 0/11 CP	GBS 0/9 CIDP 0/8 HIV enceph. 0/13 bact. mening. 6/13
Sharief	[17]	ELISA, Sandwich	2 U/ml	0/8	GBS 0/27 CIDP 07 HIV-infection 0/5

ported 5 out of 5 positive CSF from patients with HIV-1-related encephalitis [15]).

These contradictory findings cannot be explained by differences in the sensitivity of the techniques used. Paradoxically, methodologies with a lower sensitivity showed higher positivity. While patient selection and timing of serum and CSF sampling may play a crucial role in all the diseases studied, differences ranging from 0% to 100% cannot be easily interpreted, and technical problems must also be considered. Statistical comparisons between these studies were not possible.

Interleukin-2 (IL-2)

IL-2, a marker of T cell activation, is produced mainly by the Th1 subset of CD4+ lymphocytes, and acts as a growth/differentiation factor for T, B, NK and LAK cells. It is noteworthy that anti-MBP and anti-PLP T cells in MS patients have the CD4+ Th1 phenotype and function. For these reasons, IL-2 and its soluble receptor have been widely studied in both the serum and CSF of MS patients.

All but two investigations [3, 5, 18-28] described various percentages of positive sera in both RR and CP MS patients. Interestingly, the studies that obtained positive findings employed the same commercial ELISA kit having a detection limit of 0.05 U/ml, while the two negative reports used ELISA kits with a much lower sensitivity (3 U/ml). Overall, patients with active relapsing MS seem to have the highest levels of IL-2, with percentages of posivity ranging from 30% [18] to 75% [19-21].

More incongruent are the data on CP MS, where positivity ranges are from 0% [5, 17, 25] to 60% [24]. Two studies on GBS patients also reported contradictory findings: IL-2 was found in 80% of the patients in one study [23], and in none of the patients in another [5]. Moreover, conflicting findings have been reported not only in MS and GBS, but also in NIND (see Table 7). Taken all together, the IL-2 data do not indicate any clinical utility in measuring this cytokine in MS CSF and serum,

Table 7. IL-2

Authors [Ref. no.]	**Methods**	**d.l.**	**MS**		**OND**	
			serum pos/total	**CSF pos/total**	**serum pos/total**	**CSF pos/total**
Gallo [18]	ELISA, Intertest 2	0.05	6/21 acute RR 0/9 CP	9/21 acute RR 0/9 CP	vir. mening. 0/5	vir. mening. 3/5
Adachi [19]	ELISA, Intertest 2	0.05	6/7 relapsing 1/10 remitting	2/2 relapsing 0/2 remitting	n.d.	n.d.
Gallo [20]	ELISA, Intertest 2	0.05	20/36 relapsing 9/14 CP	11/36 relapsing 0/14 CP	AIDS 0/30	AIDS 0/30 n.c. 0/30
Adachi [21]	ELISA, Intertest 2	0.05	7/10 relapsing 3/17 remitting	4/6 relapsing 0/9 remitting	OND 1/10	OND 1/10
Trotter [22]	ELISA, Intertest 2	0.05	10/28 active CP 5/32 stable	n.d.	NIND 1/57 (CNS infarct)	n.d.
Gallo [23]	ELISA, Intertest 2	0.05	45/80 from 15/20 patients	18/80 from 11/20 patients	NIND 0/15	NIND 0/15
Hartung [24]	ELISA, Intertest 2	0.05	38/47 active 4/14 remitting	n.d.	GBS 34/42 CIPD 6/15 myasthenia 15/15	n.d.
Peter [3]	ELISA, Intertest 2	0.05	14/50 CP	1/50 CP	NIND 0/19	NIND 0/15
Trotter [25]	ELISA, Intertest 2	0.05	13/21 CP	n.d.	OND 2/35	n.d.
Weller [5]	ELISA, Collabor. Research	3 U/ml	0/9 RR 0/11 CP	0/9 RR 0/11 CP	GBS 0/9 CIDP 0/8 AIDS 0/13	GBS 0/9 CIDP 0/8 AIDS 0/13
Sharief [26]	ELISA, Intertest 2	0.1 U/ml	34/46 RR 0/24 CP	25/46 RR 2/24 CP	n.d.	n.d.
Ott [27]	ELISA, Biokine IL-2 TCS	3 U/ml	0/26 RR 0/5 CP	0/26 RR 0/5 CP	NIND 0/27	NIND 0/27
Sharief [28]	ELISA, Intertest 2	0.05 U/ml	31/42 relapsing 9/21 remitting	32/42 relapsing 6/21 remitting	NIND 8/26	NIND 3/26

even though patients with active relapse seem to have the highest percentages of positivity, and may have detectable IL-2 levels in their CSF. From an immunopathological point of view, the detection of IL-2 in MS CSF and serum suggests only a T cell activation in these patients.

Soluble interleukin-2 receptor (sIL-2R)

It is generally agreed that sIL-2R levels in MS serum are increased. Indeed, almost all the studies in the literature [3, 5, 19-21, 23, 26, 29, 30-36] have described increased serum levels of sIL-2R in rather large percentages of patients with active/relapsing disease, but only rarely in patients with stable/remitting disease (Table 8). CSF findings, instead, appear contradictory. While five studies failed to demonstrate detectable levels of sIL-2R in the CSF from both RR and CP active MS patients [3, 5, 20, 34, 35], five investigations by three different groups [19, 21, 26, 32, 36] reported increased sIL-2R levels in most of the patients studied (up to 100% of positivity). Increased serum levels of sIL-2R have been described in AIDS, GBS and CIDP patients, and in the CSF of patients with meningeal diseases (viral and bacterial meningitis, carcinomatous meningitis, opportunistic infections in AIDS).

Future perspectives

The detection of cytokines as soluble products in MS serum and CSF has led to conflicting results. Among the possible alternative methodologies for studying cytokine expression over the course of disease, the analysis of cytokine mRNA expression by PBL and CSF T cells by means of PCR and *in situ* hybridization techniques has recently allowed the detection of cytokine mRNA abnormalities in MS patients and in animals suffering from experimental autoimmune encephalomyelitis (EAE). In particular, increased numbers of IFNγ producing T cells have been described in the CSF of patients with active relapsing MS [37], and a relationship between TNFα mRNA expression by PBL and pre-relapse interval was recently observed in RRMS [38]. Whether these promising observations can be used for clinical and/or prognostic purposes requires futher study.

Conclusion

From a survey of the literature on cytokines in MS serum and CSF, it appears that the only noteworthy finding concerns the frequent detection of sIL-2R in the serum of active/relapsing patients. However, whether this finding can be used for clinical purposes requires further investigation. The measurement of IL-1β, TNFα, IL-2 and IL-6 in both serum and CSF, as well as the detection of sIL-2R in the CSF, seem to lack clinical utility, and cannot be taken as a marker of disease activity or progression.

Table 8. sIL-2R

Authors [Ref. no.]	**Methods**	**serum n.v.**	**MS**		**OND**	
			serum pos/total	**CSF pos/total**	**serum pos/total**	**CSF pos/total**
Greenberg [29]	ELISA	257±123	466±198	n.d.	n.d.	n.d.
Gallo [20]	ELISA, Cell-Free	185±56	22/50 active	0/50 active	AIDS 27/30	AIDS 11/30
Adachi [19]	ELISA, Cell-Free	193±29	8/10 relapsing 3/8 remitting	8/8 relapsing 0/5 remitting	n.d.	n.d.
Adachi [20]	ELISA, Cell-Free	193±29	10/13 relapsing 4/15 remitting	11/11 relapsing 0/10 remitting	n.d.	n.d.
Capra [30]	ELISA, Cell-Free	283±84	550±240 8/9 relapsing 12/25 remitting	n.d.	n.d.	n.d.
Hartung [31]	ELISA	?	30/43 relapsing 0/11 remitting	n.d.	GBS 31/50 CIDP 5/24	n.d.
Kittur [32]	ELISA, Cell-Free	372±?	470±?	14/16 stable 14/16 progr.	n.d.	vir. mening. 3/9
Bansil [33]	ELISA, Cell-Free	355±110	6/23 active CP 4/11 stable CP 0/11 ster-treat	n.d.	NIND 9/19	n.d.
Fesenmeier [34]	ELISA, Cell-Free	?	n.d.	0/24	n.d.	OND (?) 0/5
Gallo [26]	ELISA, Cell-Free	194±64	57/80 from 14 RR and 6 CP patients	7/80 from 7/14 RR and 0/6 CP patients	NIND 0/15	NIND 0/15
Peter [3]	ELISA, Cell-Free	399±140	477±359	5/50 CP	NIND 665±427	NIND 0/15
Sharief [26]	ELISA	58±31	71±25	125±32 21/46 relapsing 4/24 CP	n.d.	n.d.
Weller [5]	ELISA, Eurogenet.	187±9	229±11 RR 190±12 CP	0/9 RR 0/11 CP	HIV 381±27 GBS 225±23 CIDP 343±42	bac. men. 3/5 carci. men. 5/30
Chalon [35]	ELISA, Cell-Free	480±177	685±378	2/32	GBS 2134±2952	vir. men. 16/18 - vir. enceph. 7/7 AIDS 7/10
Sharief [36]	ELISA	?	15/42 relapsing 4/21 remitting	36/42 relapsing 13/21 remitting	NIND 4/26	NIND 0/26

From a methodological point of view, a standardization of the techniques/methods used to detect cytokines in biological fluids is urgently needed. A crucial point is the time of the lumbar puncture; for instance, in the case of TNFα assay in bacterial meningitis, this cytokine can be detected in this condition only if lumbar puncture is performed within 2-3 days of clinical onset. With regard to MS, CSF examination during acute relapse may have a greater possibility of detecting intrathecally produced IL-2 and sIL-2R. However, patient selection and classification still vary highly from group to group. The collection, handling and storage of CSF may also play a crucial role in cytokine detection. For instance, CSF should be centrifuged within a few minutes of lumbar puncture and cell counting; moreover, the addition of protease inhibitors may be necessary in the case of inflammatory and infectious diseases of the CNS, as these conditions are characterized by severe blood-brain barrier damage and/or increased cell number, and therefore by increased protease activity in the CSF. Moreover, the CSF should not be thawed more than once, i.e. just before testing, to avoid loss of biological activity.

When working with ELISA methods, the following suggestions should be considered: 1) the standard curve should be obtained by diluting known amounts of cytokine in a normal reference CSF; 2) some positive and negative CSF samples should be included in each run to test inter-test/inter-batch reproducibility; 3) if pre-coated plates are to be assayed, it must be ascertained that shipment is done correctly, and that an appropriate temperature is mantained throughout transit; 4) if an in-house ELISA is employed, check the coating and blocking steps carefully and use highly specific mAb and high-grade reagents.

As stated above, the only "acceptable" finding in all the studies cited in this review is the frequent detection of sIL-2R and, less frequently, of IL-2. From an immunopathological point of view, this strongly supports a Th1 activation in MS patients, in line with other clinical and experimental findings.

Acknowledgements. This work was supported by grants from Associazione Italiana Sclerosi Multipla (Projects 1995-1996), Ministero dell'Università e della Ricerca Scientifica (MURST 60%), and from Veneto Region.

We thank Miss Patrizia Segato for skilful assistance in preparation of the manuscript.

References

1. Hauser SL, Doolittle TH, Lincoln R, Brown RH, Dinarello CA (1990) Cytokine accumulation in the CSF in multiple sclerosis patients: frequent detection of interleukin 1 and tumor necrosis factor but no interleukin 6. Neurology 40:1735-1739
2. Maimore D, Gregory S, Arnason BGW, Reder AT (1991) Cytokine levels in the cerebrospinal fluid and serum of patients with multiple sclerosis. 32:67-74
3. Peter JB, Boctor FN, Tourtellotte WW (1991) Serum and CSF of IL-2, sIL-2R, TNFα and IL-1β in chronic progressive multiple sclerosis: expected lack of clinical utility. Neurology 41:121-123
4. Tsukada N, Miyagi K, Matsuda M, Yanagisawa N, Yone K (1991) Tumor necrosis factor and interleukin-1 in the CSF and sera of patients with multiple sclerosis. J Neurol Sci 102:230-234
5. Weller M, Stevens A, Sommer N, Melms A, Dichgans J, Wiethölter H (1991) Comparative analysis of cytokine pattern in immunological, infectious, and oncological neurological diseases. J Neurol Sci 104:215-221
6. Gallo P, Argentiero V, Piccinno MG, Giometto B, Pagni S, Bozza F, Tavolato B (1989)

Soluble mediators of the immune response in cerbrospinal fluid and serum of patients with multiple sclerosis. In: Battaglia MA (ed) Multiple Sclerosis Research. Elsevier SP, pp 61-70

7. Frei K, Leist TP, Maeger A, Gallo P, Leppert D, Zinkernagel RM, Fontana A (1988) Production of B cell stimulatory factor-2 and interferon in the central nervous system during viral meningitis and encephalitis. Evaluation in a murine model of infection and in patients. J Exp Med 168:449-453
8. Houssiau FA, Bukasa K, Sindic CJM, van Damme JV, van Snick JV (1988) Elevated levels of the 26 K human hybridoma growth factor (interleukin 6) in cerebrospinal fluid of patients with acute infection of the central nervous system. Clin Exp Immunol 71:320-323
9. Leppert D, Frei K, Gallo P, Yasargil MG, Hess K, Baumgartner G, Fontana A (1989) Brain tumors: detection of B-cell stimulatory factor-2/interleukin 6 in the absence of oligoclonal bands of immunoglobulins. J Neuroimmunol 24:259-264
10. Gallo P, Frei K, Leppert D, Fontana A (1990) On the intrathecal synthesis of immunoglobulins: detection of BSF-2/IL-6 in CSF. In: Marrosu MG, Cianchetti C, Tavolato B (eds) Trends in Neuroimmunology. Plenum Press, New York, pp 33-39
11. Paemen L, Olsson T, Söderström M, van Damme J, Opdenakkern G (1994) Evaluation of gelatinases and IL-6 in the cerebrospinal fluid of patients with optic neuritis, multiple sclerosis, and other inflammatory neurological diseases. Eur J Neurol 1:55-63
12. Frei K, Fredrikson S, Fontana A, Link H (1991) Interleukin-6 is elevated in plasma in multiple sclerosis. J Neuroimmunol 312:147-153
13. Leist EP, Frei K, Kam-Hansen S, Zinkernagel R, Fontana A (1988) Tumor necrosis factor α in cerebrospinal fluid during bacterial but not viral meningitis. J Exp Med 167:1734-1748
14. Gallo P, Piccinno MG, Krzalic L, Tavolato B (1989) Tumor necrosis factor alpha (TNFα) and neurological diseases. Failure in detecting TNFα) in the cerebrospinal fluid from patients with multiple sclerosis, AIDS dementia complex, and brain tumors. J Neuroimmunol 23:41-44
15. Franciotta DM, Grimaldi LME, Martino GV, Piccolo G, Bergamaschi R, Citterio A, Melzi d'Eril GV (1989) Tumor necrosis factor in serum and cerebrospinal fluid of patients with multiple sclerosis. Ann Neurol 26:787-789
16. Sharief MK, Noori MA, Ciardi M, Cirelli A, Thompson EJ (1993) Increased levels of circulating ICAM-1 in serum and cerebrospinal fluid of patients with active multiple sclerosis. Correlation with TNFα and blood-brain barrier damage. J Neuroimmunol 325:467-472
17. Sharief MK, Hentges R (1991) Association between tumor necrosis factor-alpha and disease progression in patients with multiple sclerosis. New Engl J Med 325:467-472
18. Gallo P, Piccinno MG, Pagni S, Tavolato B (1988) Interleukin-2 levels in serum and cerebrospinal fluid of multiple sclerosis patients. Ann Neurol 24:795-797
19. Adachi K, Kumamoto T, Araki S (1989) Interleukin-2 receptor levels indicating relapse in multiple sclerosis (Letter). Lancet i:559-560
20. Gallo P, Piccinno MG, Pagni S, Argentiero V, Giometto B, Bozza F, Tavolato B (1989) Immune activation in multiple sclerosis: study of IL-2, sIL-2R, and IFNγ in serum and cerebrospinal fluid. J Neurol Sci 92:9-15
21. Adachi K, Kumamoto T, Araki S (1990) Elevated soluble interleukin-2 receptor levels in patients with active multiple sclerosis. Ann Neurol 28:687-691
22. Trotter JL, van der Veen RC, Clifford DB (1990) Serial studies of interleukin-2 in chronic progressive multiple sclerosis: occurence of "bursts" and effect of cyclosporine. J Neuroimmunol 28:9-14
23. Gallo P, Piccinno MG, Tavolato B, Sidèn Å (1991) A longitudinal study on IL-2, sIL-2R, IL-4 and IFNγ in multiple sclerosis CSF and serum. J Neurol Sci 101:227-232

24. Hartung HP, Reiners K, Schmidt B, Stoll G, Toyka KV (1991) Serum interleukin-2 concentrations in Guillain-Barré syndrome and chronic idiopathic demyelinating polyradiculoneuropathy: comparison with other neurological diseases of presumed immunopathogenesis. Ann Neurol 30:48-53
25. Trotter JL, Collins KG, van der Veen RC (1991) Serum cytokine levels in chronic progressive multiple sclerosis: interleukin-2 levels parallel tumor necrosis factor-α levels. J Neuroimmunol 33: 29-36
26. Sharief MK, Hentges R, Thompson EJ (1991) The relationship of interleukin-2 and soluble interleukin-2 receptors to intrathecal immunoglobulin synthesis in patients with multiple sclerosis. J Neuroimmunol 32:43-51
27. Ott M, Demisch L, Engelhardt W, Fisher PA (1993) Interleukin-2, soluble interleukin-2 receptor, neopterin, L-tryptophan, and $\beta_2$microglobulin levels in CSF and serum of patients with relapsing-remitting or chronic-progressive multiple sclerosis. J Neurol 241:108-114
28. Sharief MK, Thompson EJ (1993) Correlation of interleukin-2 and soluble interleukin-2 receptors with clinical activity of multiple sclerosis. J Neurol Neurosurg Psychiatry 56:169-174
29. Greenberg SJ, Marcon L, Hurwitz BJ, Waldmann TA, Nelson DL (1988) Elevated levels of soluble interleukin-2 receptors in multiple sclerosis (letter). N Engl J Med 319:1206
30. Capra R, Mattioli F, Marciano N, Vignolo LA, Bettinzioli M, Airò P, Cattaneo R (1990) Significantly higher levels of soluble interleukin-2 in patients with relapsing-remitting multiple sclerosis compared to healthy subjects (lettere). Arch Neurol 47:254
31. Hartung HP, Hughes RAC, Taylor WA, Heininger K, Reiners K, Toyka KVT (1990) T cell activation in Guillain-Barré syndrome and in MS: elevated serum levels of IL-2 receptors. Neurology 40:215-218
32. Kittur SD, Kittur DS, Soncrant TT, Rapoport SI, Tourtellotte WW, Nagel JL, Adler WH (1990) Soluble interleukin-2 receptors in cerebrospinal fluid from individuals with various neurological disorders. Ann Neurol 28:169-173
33. Bansil S, Troiano R, Cook SD, Rohowsky-Kochan C (1991) Serum soluble interleukin-2 receptor levels in chronic progressive, stable, and steroid treated multiple sclerosis. Acta Neurol Scand 84:282-285
34. Fesenmeier JT, Whitaker JN, Herman PK, Walker DP (1991) Cerebrospinal fluid levels of myelin basic protein-like material and soluble interleukin-2 receptor in multiple sclerosis. J Neuroimmunol 34:77-80
35. Chalon MP, Sindic CJM, Laterre EC (1993) Serum and CSF levels of soluble interleukin-2 receptors in MS and other neurological diseases: a reappraisal. Acta Neurol Scand 87:77-82
36. Sharief MK, Hentges R, Ciardi M, Thompson EJ (1993) In vivo relationship of interleukin-2 and soluble interleukin-2 receptor with clinical activity in multiple sclerosis. J Neurol 240:46-50
37. Link J, Söderström M, Olsson T, Höjeberg B, Ljungdahl Å, Gustafsson A, Link H. Increased TGFβ, IL-4 and IFNγ in multiple sclerosis. Ann Neurol 36:379-386
38. Rieckmann P, Albrecht BA, Kitze B, Weber T, Tumani H, Broocks A, Luer W, Helwig A, Poser S (1995) Tumor necrosis factor-α messenger RNA expression in patients with relapsing-remitting multiple sclerosis is associated with disease activity. Ann Neurol 37:82-88

Cerebrospinal fluid markers of demyelination: MBP and anti-brain protein antibodies

P. Annunziata

MBP and demyelination in multiple sclerosis

In multiple sclerosis (MS), an inflammatory mechanism leads to demyelination and release of myelin proteins. The destroyed myelin is engulfed by macrophages as vesicular material. One of the most studied proteins is myelin basic protein (MBP), which constitutes about 30% of the protein component of central nervous system myelin: it is capable of inducing experimental allergic encephalomyelitis (EAE), an animal model close to MS pathology. MBP has been detected in the cerebrospinal fluid (CSF) of patients with MS during acute exacerbation up to two weeks after clinical onset, and is considered a reliable indicator of acute demyelination [1-3]. The MBP-like material detected in MS CSF has been extensively studied. By means of antisera recognizing different epitopes of the molecule, it was found that MBP circulating in CSF is a fragment containing an epitope corresponding to aminoacid residues 45-89 and that peptide 80-89 is the smallest sequence containing the dominant epitope [4]. The MBP epitope detected in CSF was found to be different from that in the urine of MS patients which corresponds to peptide 82-89 [5]. These findings suggest that MBP fragments released during myelin breakdown undergo enzymatic modification in the blood and additionally in the kidneys. They also contribute to understanding of the mechanism underlying MBP processing and how humoral and cellular immune responses directed against myelin proteins are elicited.

The molecular and immunological mechanism underlying the demyelination process in MS is unclear. Currently, the results of *in vitro* experiments have provided clues to the demyelination process *in vivo*. Controversy exists as to the role of humoral immune responses against myelin components. Immunocytochemical studies have shown the presence of B cells with complement activation fractions in acute MS lesions and macrophages in contact with myelin have been found to contain immunoglobulin-complement complexes [6].

A number of antimyelin proteins and oligodendrocyte antibodies have been reported in CSF of patients with MS. A comprehensive list of the myelin proteins involved is shown in Table 1.

Institute of Neurological Sciences, University of Siena, V.le Bracci 2, 53100 Siena, Italy

Table 1. Brain proteins involved in intrathecal B cell response in MS

Protein	Molecular weight (kD)	Location	Antibody isotype
MBP	18.5	Internal CNS myelin sheat	IgG, IgM
PLP	24	CNS myelin	IgG
MAG	72-67	CNS and PNS myelin	IgG, IgM
MOG	54-27	Oligodendrocyte membrane	IgG
CSL	75-55-43-33-31.5	Oligodendrocytes Schwann cells, neurons, astrocytes, ependymal cells	IgG

MBP = myelin basic protein; PLP = proteolipid protein; MAG = myelin-associated glycoprotein; MOG = myelin-oligodendrocyte glycoprotein; CSL = cerebellar soluble lectin; CNS = central nervous system; PNS = peripheral nervous system.

Antioligodendrocyte antibodies

Despite conflicting results in sera due to non specific binding to brain tissue or cells, mediated by Fc-receptors, antioligodendrocyte antibodies have been detected in some CSF samples from patients with MS by sensitive techniques that avoid the problem of false binding specificity. These antibodies have also been found in other inflammatory and non-inflammatory neurological diseases [7-9]. By contrast, some Authors, employing unconcentrated CSF from MS patients, to avoid non specific binding to oligodendrocyte cultures, failed to detect any antibody binding to rat or human oligodendrocytes [10].

These findings excluded a specific pathogenetic role *in vivo* of these antibodies. However, recent data *in vitro* suggest that they have an indirect role in determining cytotoxicity. In fact, in co-cultures of oligodendroglia and macrophages, destruction of oligodendrocytes occurred only in the presence of antibodies directed against their surface components. These antibodies stimulated macrophage attachment and destruction of oligodendroglial cells [11].

Antimyelin basic protein antibodies

MBP has been extensively studied because of its ability to cause experimental allergic encephalomyelitis (EAE), the animal model most similar to MS. The search for antibodies directed against this protein has therefore received much attention but has yielded conflicting results due to methodological problems, the principal of which is due to the strong positive charge of MBP that induces non specific IgG binding. By using methods such as complement fixation assay, radioimmunoassay, ELISA and immunoblotting, anti-MBP IgG was found in CSF samples from patients with MS by some Authors but not by others. For instance, Chou et al. [12] failed to detect any CSF specific antibody directed against whole human MBP and three peptic fragments of the molecule by sensitive solid-phase radioimmunoassay. It should be mentioned that the CSF samples were obtained from 10 patients with chronic progressive MS. However, in nearly 30% of another group of 37 patients with MS anti-MBP IgG were detected by an ELISA method [13]. These antibodies belonged to the IgG_1 or IgG_3 subclasses, suggesting a monoclonal or, at least in some cases, oligoclonal origin. The oligoclonal nature of CSF anti-MBP IgG in some MS patients was confirmed by sensitive immunoblot transfer from agarose isoelectric focusing [14]. These specific bands did not comigrate with the typical CSF oligoclonal IgG bands. However, some of the patients found to have anti-MBP antibodies by the immunoblot assay were not antibody-positive to ELISA, suggesting that this method has low sensitivity.

Further incentive to investigate humoral immune response directed against MBP was provided by the detection of immune complexes containing MBP in CSF. A different pattern of anti-MBP antibodies, free or bound in immune complexes, was found in relation to disease activity. High levels of free anti-MBP antibodies were detected in MS patients during exacerbations, while in chronic progressive patients these antibodies were predominantly in the bound form; no detectable anti-MBP antibodies were found in patients in clinical remission [15]. Furthermore, free anti-MBP antibodies correlated with free CSF MBP levels in patients with active disease.

Longitudinal studies in some patients showed high titres of free anti-MBP antibodies during acute exacerbations and high bound levels during progressive stage of disease [15]. Free anti-MBP antibody levels in CSF of patients during exacerbation were found to be neutralized by CSF samples from patients in remission; this neutralization was inhibited when CSF from chronic progressive patients was added [16]. These results support the hypothesis of the existence of an idiotypic network controlling anti-MBP antibody function. All these studies suggest that anti-MBP antibodies are immunological factors involved in the pathogenesis of demyelination in active MS, and not merely a secondary response to myelin destruction.

The failure to detect anti-MBP antibodies in some MS patients could be due to binding to target antigens and subsequent catabolism, leading to low undetectably levels of circulating antibody. This hypothesis is supported by experimental data in chronic relapsing EAE, in which antibodies directed against MBP were detected *in vitro* in cultures of mononuclear cells isolated from the central nervous system but not in brain extracts [17]. MBP-specific antibody-forming B cells were recently found in *post mortem* brain tissue of MS victims, but not victims of other neurological and non-neurological diseases [18]. These findings focus on the possible role of anti-MBP antibodies *in situ* in central nervous system demyelination. To detect antibody production by cells independently of circulating levels, a sensitive solid-phase enzyme linked immunospot assay was employed. Cells producing anti-MBP IgG antibodies were found in CSF, but not in peripheral blood, of 57% of MS patients . These cells were rarely detected in CSF of patients with acute aseptic meningoencephalitis [19]. This data confirms the existence of an important intrathecal B cell response directed against myelin and specifically MBP. It raises the possibility of local anti-MBP antibody production even in the absence of detectable circulating levels in CSF.

In the last few years, new insights into the immune response against MBP emerged from studying the epitope specificity of these antibodies. Anti-MBP IgG antibodies purified from CSF of MS patients during acute exacerbation were found to be specific for the epitope of MBP synthetic peptides corresponding to aminoacid residues 61-106 of the molecule. In addition, anti-MBP IgG isolated from soluble extracts of several areas of an autopsy brain tissue of an MS victim showed similar epitope specificity to IgG isolated from CSF, whereas tissue-bound anti-MBP IgG had more restricted specificity suggesting that the exact epitope recognized is located in the range 84-95 of MBP aminoacid sequence [20]. These results suggest that not all anti-MBP antibodies represent a secondary response to demyelination, but selected MBP epitopes may be recognized by specific antibodies in some brain areas and cause demyelination by complement fixation or macrophage recruitment and activation.

Antimyelin basic protein IgM

The many studies on humoral immunity directed against MBP have been concerned predominantly with the IgG isotype. The role of the anti-MBP IgM response in MS has not yet been studied, although total IgM in CSF has been measured. To test whether an important primary response against MBP could take place in MS patients and could be related to prognostic evolution of disease, we recently assayed CSF IgM binding to human MBP by ELISA in patients with relapsing-remitting MS. We found elevated intrathecal IgM anti-MBP levels in nearly 30% of patients and this response was significantly associated with a low disease progression rate [21]. In addition, this patient subgroup with elevated anti-MBP IgM levels showed a less prominent magnetic resonance imaging (MRI) pattern than those with low or absent IgM response. These findings, currently under further investigation in our laboratory, provide additional evidence for a possible role of the humoral primary response directed against MBP in regulating demyelination process in MS. How this role could be explicated and whether other humoral or cellular factors are involved in this process, remains matter of further investigation.

Antiproteolipid protein antibodies

The absence of CSF anti-MBP antibodies in some MS patients raised the possibility of detecting immunoglobulins directed against other important myelin components. Proteolipid protein (PLP) is the main candidate because it constitutes more than 50% of myelin proteins and was found to be able to induce EAE. In a large series of 385 MS patients, anti-PLP IgG was found in CSF of seven patients with clinical relapse, four with the progressive form and none in clinical remission [22]. The majority of the patients had anti-MBP antibodies, and high levels of both antibodies were never detected in the same subject. Autopsy of four MS patients showed concordance between the CSF antibody pattern and IgG isolated from brain tissue;

furthermore, the brain of one case with tissue anti-PLP IgG showed less abundant inflammation than those with anti-MBP IgG. These findings, although not yet confirmed by other Authors, focus on the possible existence in MS of a different pattern of immune mechanism leading to demyelination in which B cell responses directed against MBP or PLP may play an important role. Furthermore, anti-PLP IgG-secreting cells were found in CSF from patients with MS at higher number than corresponding peripheral blood cells and than in patients with aseptic meningitis or other neurological diseases [23]. Recently, anti-PLP IgG-secreting cells were also found in the majority of patients with idiopathic optic neuritis, and with optic neuritis as a symptom of MS, but in a small percentage of other inflammatory neurological diseases [24].

Antimyelin-associated glycoprotein and anti-myelin-oligodendrocyte glycoprotein antibodies

The finding of loss of myelin-associated glycoprotein (MAG) in active MS plaques stimulated the search for an immune response directed against MAG, another important constituent of central myelin. In one study, anti-MAG antibodies were detected in some CSF samples but not in peripheral blood of MS patients. Anti-MAG IgG-secreting cells were found in CSF of about 50% of untreated MS patients but less frequently in parallel peripheral blood samples. Only 2 patients showed CSF IgG-secreting cells directed against MAG and MBP [25]. The role of MAG as putative antigen triggering autoimmune process in MS remains to be demonstrated. However, ultrastructural changes in central myelin have been found *in vivo* after injection of monoclonal antibodies to MAG.

Another potential antigen target of intrathecal response against myelin investigated in MS is myelin-oligodendrocyte glycoprotein (MOG), a minor component of CNS myelin, located at the myelin membrane surface. Interest in this membrane glycoprotein is also due to the finding that an anti-MOG monoclonal antibody induced acceleration of clinical symptoms and demyelination pattern similar to those of MS in animals with MBP-induced EAE [26]. Anti-MOG IgG-secreting cells were detected in most CSF samples from MS patients at higher number than that found in the peripheral blood compartment. Some patients showed anti-MOG IgG-secreting cells only in the CSF but not in peripheral blood [27].

Anticerebellar soluble lectin antibodies

An endogenous mannose-binding cerebellar soluble lectin (CSL) has been found in oligodendrocytes, neurons, astrocytes and tight junctions of ependymal cells. This glycoprotein stabilizes the myelin structure by forming a molecular bridge with the carbohydrate moiety of myelin glycoproteins. Anti-CSL antibodies were detected by immunoblotting in the CSF from most of a large population of patients with MS. They were also detected with a lower incidence in other inflammatory and non-inflammatory neurological diseases, predominantly in aged patients [28]. It was proposed that in MS patients under 50 years of age CSF anti-CSL antibodies are a sensitive and specific diagnostic test. The presence of these antibodies could be related to changes in the ependymal layer around the cerebral ventricles in MS patients and to subsequent demyelination, as indicated by *in vitro* experiments with anti-CSL antibody. Further investigation is needed to establish whether CSL could be a novel antigen target triggering autoimmune responses against the central nervous system in MS.

The role of antibrain proteins in MS demyelination

In the last decade, a large number of studies have confirmed the existence of intrathecal B cell responses directed against several myelin proteins and glycoproteins in patients with MS. Despite of lack of disease-specificity, this response showed two important features, predominantly highlighted by elispot assay studies: 1) it is more prevalent in CSF than in peripheral blood; 2) it may be directed simultaneously against more than one antigen target in the same patient.

The intrathecal compartmentalization of this humoral response is in line with current knowledge on the origin and extent of the inflammatory mechanism in MS characterized by peripheral blood mononuclear cell recruitment and subsequent transfer across the blood-brain barrier. The demonstration of a "monospecific" B cell response directed against one predominant myelin protein is more intriguing, raising the hypothesis that a "personal" autoimmune response could take place in each MS patient. Whether this predominant response initiates or regulates demyelination process during subsequent relapses is uncertain; however, a parallel T cell response could be just as significant. The induction of a "chimeric" experimental model, with transfer of these B cell clones from man to rat, could provide answers to these questions.

References

1. Cohen SR, Herndon RM, McKhann GM (1976) Myelin basic protein in cerebrospinal fluid as an indicator of active demyelination. Radioimmunoassay of myelin basic protein in spinal fluid. An index of active demyelination. N Engl J Med 295:1454-1457
2. Whitaker JN (1977) Myelin encephalitogenic protein fragments in cerebrospinal fluid of persons with multiple sclerosis. Neurology 27:911-920
3. Whitaker JN, Lisak RP, Bashir RM et al (1980) Immunoreactive myelin basic protein in the cerebrospinal fluid in neurological disorders. Ann Neurol 7:58-64
4. Whitaker JN, Gupta M, Smith OF (1986) Epitopes of immunoreactive myelin basic protein in human cerebrospinal fluid. Ann Neurol 20:329-336
5. Whitaker JN (1987) The presence of immunoreactive myelin basic protein peptide in urine of persons with multiple sclerosis. Ann Neurol 22:648-655
6. Prineas JW, Graham JS (1981) Multiple sclerosis: capping of surface immunoglobulin G on macrophages engaged in myelin breakdown. Ann Neurol 10:149-158

7. Traugott U, Raine CS (1981) Antioligodendrocyte antibodies in cerebrospinal fluid of multiple sclerosis and other neurologic diseases. Neurology 31:695-700
8. Gorny MK, Wroblewska Z, Pleasure D, Miller SL, Wajgt A, Koprowski H (1983) CSF antibodies to myelin basic protein and oligodendrocytes in multiple sclerosis and other neurological diseases. Acta Neurol Scand 67:338-347
9. Steck A, Link H (1984) Antibodies against oligodendrocytes in serum and CSF in multiple sclerosis and other neurological diseases: ^{125}I-protein A studies. Acta Neurol Scand 69:81-89
10. Lubetzki C, Lombrail P, Hauw JJ, Zalc B (1986) Multiple sclerosis: rat and human oligodendrocytes are not the target for CSF immunoglobulins. Neurology 36:524-528
11. Scolding NJ, Compston DA (1991) Oligodendrocyte-macrophage interactions in vitro triggered by specific antibodies. Immunology 72:127-132
12. Chou C-HJ, Tourtellotte W, Kibler RF (1983) Failure to detect antibodies to myelin basic protein or peptic fragments of myelin basic protein in CSF of patients with MS. Neurology 33:24-28
13. Garcia-Merino A, Persson MAA, Ernerudh J, Diaz-Gil JJ, Olsson T (1986) Serum and cerebrospinal fluid antibodies against myelin basic protein and their IgG subclass distribution in multiple sclerosis. J Neurol Neurosurg Psychiatry 49:1066-1070
14. Cruz M, Olsson T, Ernerudh J, Hojeberg B, Link H (1987) Immunoblot detection of oligoclonal anti-myelin basic protein IgG antibodies in cerebrospinal fluid in multiple sclerosis. Neurology 37:1515-1519
15. Warren K, Catz I (1987) A correlation between cerebrospinal fluid myelin basic protein and anti-myelin basic protein in multiple sclerosis patients. Ann Neurol 21:183-189
16. Warren KG, Catz I (1988) Neutralization of anti-myelin basic protein by cerebrospinal fluid of multiple sclerosis patients in clinical remission. J Neurol Sci 88:185-194
17. Olsson T, Henriksson A, Link H (1985) In vitro synthesis of immunoglobulins and autoantibodies by lymphocytes from various body compartments during chronic-relapsing experimental allergic encephalomyelitis. J Neuroimmunol 9:293-305
18. Gerritse K, Deen C, Fasbender M, Ravid R, Boersma W, Claassen E (1994) The involvement of specific anti-myelin basic protein antibody-forming cells in multiple sclerosis immunopathology. J Neuroimmunol 49:153-159
19. Olsson T, Baig S, Hojeberg B, Link H (1990) Antimyelin basic protein and antimyelin antibody-producing cells in multiple sclerosis. Ann Neurol 27:132-136
20. Warren KG, Catz I (1993) Increased synthetic peptide specificity of tissue-CSF bound anti-MBP in multiple sclerosis. J Neuroimmunol 43:87-96
21. Annunziata P, Martino T, Maimone D, Guazzi GC (1994) Intrathecal IgM anti-MBP levels and prognosis in multiple sclerosis. J Neurol (Suppl 1) 241:S152
22. Warren KG, Catz I, Johnson E, Mielke B (1994) Anti-myelin basic protein and anti-proteolipid protein specific forms of multiple sclerosis. Ann Neurol 35:280-289
23. Sun J-B, Olsson T, Wang W-Z, Xiao B-G, Kostulas V, Fredrikson S, Ekre H-P, Link H (1991) Autoreactive T and B cells responding to myelin proteolipid protein in multiple sclerosis and controls. Eur J Immunol 21:1461-1468
24. Sellebjerg FT, Fredriksen JL, Olsson T (1994) Anti-myelin basic protein and anti-proteolipid protein antibody-secreting cells in the cerebrospinal fluid of patients with acute optic neuritis. Arch Neurol 51:1032-1036
25. Baig S, Olsson T, Yu-Ping J, Hojeberg B, Cruz M, Link H (1991) Multiple sclerosis: cells secreting antibodies against myelin-associated glycoprotein are present in cerebrospinal fluid. Scand J Immunol 33:73-79
26. Linington C, Bradl M, Lassmann H, Brunner C, Vass K (1988) Augmentation of demyelination in rat acute allergic encephalomyelitis by circulating mouse monoclonal antibodies directed against a myelin-oligodendrocyte glycoprotein. Am J Pathol 130:443-454

27. Sun J, Link H, Olsson T, Xiao BG, Andersson G, Ekre HP, Linington C, Diener P (1991) T and B cell responses to myelin-oligodendrocyte glycoprotein in multiple sclerosis. J Immunol 146:1490-1495
28. Zanetta J-P, Tranchant C, Kuchler-Bopp S, Lehmann S, Warter JM (1994) Presence of anti-CSL antibodies in the cerebrospinal fluid of patients: a sensitive and specific test in the diagnosis of multiple sclerosis. J Neuroimmunol 52:175-182

Effects of immunotherapeutic strategies on cerebrospinal fluid parameters in multiple sclerosis

K.J.B. Lamers[1], S.T.F.M. Frequin[2] and O.R. Hommes[1]

Introduction

Therapeutic strategies in MS are generally based on treatment with anti-inflammatory, immunosuppressive or immunomodulatory agents. Most applied therapeutic agents in MS are: corticosteroids, cyclophosphamide, azathioprine, cyclosporine, interferon, total lymphoid irradiation, plasma exchange, monoclonal antibodies, copolymer-I and immunoglobulins. Modes of administration are: oral, intramuscular, intravenous or intrathecal; administration of drugs can be carried out by both a short course of high dose or a chronic low dose during an extended period.

Effects of therapeutic strategies in MS on clinical variables and CSF parameters can be measured at different time intervals. Both short term and long term effects are reported. It can be expected that these differences in 1) therapeutic agents, 2) modes of administration, 3) duration and quantity of drug administration, 4) disease course of patients, and 5) evaluation periods, influence the effects on CSF parameters. Therefore, in studying the effects of immunotherapeutic strategies on CSF parameters one has to keep in mind that the results of the various studies are difficult to compare. In this overview, 4 groups of relevant CSF parameters have been selected: a) humoral immunity, b) cellular immunity, c) blood/CSF barrier, d) myelin basic protein. The effects of treatment on these CSF variables have been studied and described.

Humoral immunity

IgG index and intrathecal IgG production

Several reports have been published with respect to the effects of therapeutic approaches in MS patients on intrathecally produced IgG. Intrathecally produced IgG can be established by using 1) IgG index, 2) formulas for calculating intrathecal IgG, and 3) IgG oligoclonality by isoelectric focusing. In a study by Frequin [1], 101

[1] University Hospital Nijmegen, Institute of Neurology, PO Box 9101, 6500 HB Nijmegen, The Netherlands; [2] St. Antonius Hospital, Institute of Neurology, PO Box 2500, 3430 EM Nieuwegein, The Netherlands

patients with relapsing-remitting (RR) or chronic progressive (CP) MS were treated with 1000 mg intravenous methylprednisolone (IVMP) for 10 consecutive days. All patients had a symptomatic deterioration of the disease. Just before and immediately after IVMP treatment, CSF intrathecal IgG production was measured; the mean intrathecal IgG production had reduced from 31.0 mg/L to 11.6 mg/L (63%) (Table 1). There was no correlation between the reduction in intrathecal IgG in individual patients and the reduction in EDSS score.

38 patients were also included in a longitudinal prospective study (Frequin et al. [2]). The mean follow-up period after entry was 2.6 years; after this period the mean intrathecal IgG synthesis had decreased from 27.5 to 13.9 mg/L (50%) (Table 2). This result demonstrates that more than two years after IVMP treatment intrathecal IgG is still reduced while blood IgG level is almost restored.

Anderson et al. [3] treated 26 MS patients with IVMP (1 g daily for 7 days), and CSF and blood were taken immediately prior to and 14 days following therapy. The daily CSF IgG synthesis rate had markedly decreased from 37.1 to 15.6 mg per day (60%); no correlation could be shown between clinical improvement and the degree of reduction in IgG synthesis.

Wender et al. [4] compared the IgG index reduction in 3 therapeutical trials: they found that treatment with high-doses of prednisone was more effective in suppressing CNS IgG synthesis than treatment with ACTH or cyclophosphamide.

Table 1. Mean and median values of clinical scoring (EDSS) and CSF parameters in 101 MS patients before and after treatment with high dose intravenous methylprednisolone (MP) (1000 mg daily for 10 consecutive days)

	Ref. values		**Unit**	**MS total (n=101 patients)**						**Sign**
				Pre MP			**Post MP**			
	mean	P_{90}		mean	median	$>P_{90}$ in%	mean	median	$>P_{90}$ in%	
EDSS	-	-	-	4.4	4.0	-	3.3	3.0	-	*
Q Albumin x 10^{-5}	527	745	-	637	499	20	518	480	21	ns
CSF monon. cells	1	7	µl	6	2	48	3	2	44	*
CSF IgG	23.9	43.0	mg/l	62.5	46.0	59	36.1	29.5	24	*
IgG index	0.47	0.54	-	1.05	0.87	94	0.98	0.74	84	*
Intrathecal IgG	0	0	mg/l	31.0	14.0	85	11.6	4.0	70	*
CSF IgM	0.3	0.4	mg/l	2.1	0.6	66	0.9	0.5	56	*
IgM index	0.05	0.1	-	0.13	0.06	31	0.12	0.06	31	ns
Intrathecal IgM	0	0	mg/l	1.0	0	32	0.4	0	28	*
IEF	0	0	-	5	6	92	4	3	82	*
CSF MBP	0.6	1.2	µg/l	1.6	0.8	33	0.7	0.5	8	*

EDSS = expanded disability status scale; Q Albumin = CSF Alb/Serum Alb; CSF monon. cells = number of mononuclear cells; IEF = iso-electric focusing, number of oligoclonal bands; CSF MBP = myelin basic protein; % >P90 = percentage of MS patients with CSF value higher than P_{90} of the reference value; Sign = significance between pre and post MP treatment; ns = not significant; * = $p<.01$

Table 2. Results of CSF variables in 38 MS patients before and after treatment with high-dose intravenous methylprednisolone (IVMP) and after the follow-up period (2.6 years)

	Ref. value	**Unit**	**Before IVMP**	**After IVMP**		**Follow-up**	
	mean		mean	mean	p value**	mean	p value***
Mononuclear cells	1	/μl	6.7*	3	0.01	5*	ns
MBP	0.6	μg/L	2.1*	0.6	0.0003	0.4*	0.0001
IEF	0	-	5.4*	3.7*	0.0001	4.2*	ns
Q Albumin (x 10^{-5})	527	-	579	531	ns	534	ns
IgG	23.9	mg/L	62.9*	33.8*	0.0000	41.2*	0.004
IgG index	0.5	-	1.0*	0.9*	0.002	0.9*	ns
Intrathecal IgG	0	mg/L	27.5*	9.2*	0.0000	13.9*	ns

MBP = myelin basic protein; IEF = iso-electric focusing number of oligoclonal bands; Q = quotient CSF/serum; ns = not significant; * = $p<0.05$ significant different with respect to reference values; ** = p value significant different with respect to Before IVMP; *** = p value significant different with respect to After IVMP.
p values were calculated using Wilcoxon's signed rank test.

They did not observe any correlation between depression of intrathecal IgG synthesis and the clinical outcome.

Durelli et al. [5] conducted a double-blind trial of high-dose parenteral 6-MP and placebo on 23 patients with MS. After the trial, patients were given corticosteroids in gradually decreasing doses; 3 to 6 days after starting parenteral MP, the CNS IgG synthesis rate was markedly decreased, on average 10% of the pre-treatment value. However, in contrast to the results of Frequin et al. [1], the CNS IgG synthesis rate increased after 60 days to 80-100% of the pre-treatment value. There was no correlation between CNS IgG production rate decrease and corresponding clinical improvement.

Warren et al. [6] treated 40 active RR patients (placebo, low-dose IV, ACTH, high-dose IVMP and mega-dose IVMP). After 10 days of treatment, intrathecal IgG had not changed in placebo and ACTH patients, while intrathecal IgG had dramatically decreased in IVMP patients. Among immunosuppressants, azathioprine (AZA) is widely used in clinical trials; in a trial with 66 relapsing MS patients, 40 patients received each 2.5 mg/kg/day AZA, and all patients received a course of dexamethasone (DEXA) during clinical relapses. After 2 years, the mean IgG index had clearly decreased from 1.64 to 1.04 in the group of AZA treated patients, while the IgG index had not changed in the group of patients treated with DEXA (Caputo et al. [7]). The reduction in IgG index was more significant in patients with disease duration of less than 3 years, disability status of less than 3 years, and IgG index higher than 1.5. The latter finding is supported by the high correlation between the pre-treatment value of IgG index and its reduction after AZA therapy.

The finding that AZA plays a role in reducing intrathecal IgG synthesis is in contrast with the observations of another group, which demonstrated that the effect of a combined administration of AZA and steroids upon the IgG synthesis rate was similar to that of steroids, suggesting that AZA and steroids in combination were not more effective in reducing intrathecal IgG synthesis than steroids alone (Stangaitis et al. [8]).

In a study by Hommes et al. [9], 40 chronic progressive MS patients had an intensive immunosuppressive treatment with 400 mg cyclophosphamide and 100 mg prednisolone per day, with a total dose of 8 g cyclophosphamide. Immediately after treatment, both serum IgG and intrathecally produced IgG were reduced to about 50% and 60% of the starting value, respectively. After a three months period, the serum levels returned to normal but intrathecal CSF IgG values remained low. In 9 patients the IgG levels were determined again after 26 months and intrathecal IgG still showed the same low levels as immediately after treatment. Furthermore, patients with high IgG index before treatment had a good prognosis after intensive immunosuppression.

Wajgt et al. [10] treated in total 105 patients by using 4 therapeutic regimens: mega-dose prednisone, moderate-dose prednisone, mega-dose solumedrol, and intravenous cyclophosphamide. All these therapeutic regimens had a significant reducing effect on IgG index.

Panitch et al. [11] treated 12 RR MS patients in a placebo-controlled, double-blind protocol by daily subcutaneous injection of IFN-α. During the 6 months treatment, serum IgG increased dramatically: in 6 patients the mean IgG index levels had increased by 75% and increased indices were associated with positive clinical responses, while the 6 patients with unchanged or decreased IgG indices still showed moderate or severe exacerbations on IFN-α.

In conclusion, immunotherapeutic agents like corticosteroids and immunosuppressants in MS cause a clear reduction in intrathecal IgG; the effect lasts for at least 3 months, and mostly a long term effect (2 years) is noticeable. Serum IgG levels restore rapidly; the reduction in intrathecal IgG level is not related to clinical improvement, and IFN-α treatment seems to increase IgG levels.

Oligoclonal IgG bands

Oligoclonal bands in CSF can be detected by electrophoresis or by isoelectric focusing (IEF). Bands can be visualized by protein staining or after immunoblotting with subsequent immunoperoxidase labelling.

Frequin et al. [1] reported on the effect of high-dose IVMP treatment in 101 MS patients; IEF of CSF and serum was performed with LKB Ampholine PAG-plates and Coomassie blue staining. The median number of CSF oligoclonal bands had reduced from 6 to 3 bands after the treatment (Table 1); in 10 patients the oligoclonal pattern had completely disappeared after treatment. In a follow-up study of 38 patients CSF analyses were repeated; 2.6 years after start of therapy the mean number of CSF bands was still reduced from 5.4 to 4.2 (Table 2).

Durelli et al. [5] treated 23 MS patients in a double-blind trial with high-dose parenteral 6-MP and placebo. CSF IgG oligoclonal bands were assessed by means of cellulose acetate electrophoresis and protein staining. Despite the pronounced decrease in CNS IgG synthesis rate, the number of bands was unchanged after 3 to 15 days of parenteral MP, but the intensity of oligoclonal banding was reduced.

Conversely, on day 60 of treatment the CSF bands had disappeared in 1 of the 4 patients in whom CSF was studied, and in 1 patient 3 bands were reduced to 1 band.

Baumhefner et al. [12] treated 9 chronic progressive MS patients with high-dose IVMP; oligoclonal bands were detected by IEF, immunofixation and silver staining. After treatment there was no change in intensity or pattern of IgG bands. Anderson et al. [3] treated 26 patients with IVMP (1 g daily for 7 days). Oligoclonal bands were determined by agarose gel electrophoresis, immunofixation and protein staining; oligoclonal bands disappeared following treatment in 3 patients, and became equivocal in 4.

Confavreux et al. [13] treated 11 MS patients intrathecally with natural ß-interferon (IFN-ß) (100.000 U at weekly intervals for 8 weeks). IEF was performed on Ampholine PAG-plates, followed by silver staining; the banding pattern was remarkably stable within any individual over a 6 months period. Among immunosuppressants, treatment of MS patients with AZA and DEXA (Caputo et al. [7]), or AZA and prednisone (Stangaites et al. [8]), did not show any changes in CSF oligoclonal banding pattern. In a therapeutic trial on 49 MS patients with an intensive immunosuppressive treatment with cyclophosphamide and prednisone no significant change in CSF band pattern could be observed 3 months after start of therapy (Van Munster [14]).

In conclusion, IVMP treatment has a short and a long term reducing effect on the number of CSF bands. In some patients IgG oligoclonality disappears; however, some studies do not find these changes in oligoclonality. This effect cannot be demonstrated in case of treatment with high-dose immunosuppressants, like AZA or cyclophosphamide, or in case of treatment with IFN-ß. It is remarkable that although intrathecal IgG production is reduced by both treatment regimens (IVMP and cyclophosphamide), oligoclonal IgG bands are only reduced by IVMP treatment.

IgM index, intrathecal IgM and oligoclonal IgM bands

Intrathecal IgM production has been found in 30-60% of MS patients by quantitative and/or qualitative assays. The recommended method for qualitative detection of oligoclonal IgM bands is electrophoresis, or IEF (agarose) of unconcentrated CSF and subsequent immunodetection. Sharief et al. [15] demonstrated that intrathecal production of IgM in MS patients was correlated with disease activity, manifesting as a recent relapse, as well as with the total number of relapses. Frequin et al. [1] reported that the mean CSF intrathecal IgM production was high in active RR patients (2.8 mg/L) and normal in deteriorating CP patients (0.3 mg/L).

In RR patients, intrathecal IgM and not intrathecal IgG was related with CSF myelin basic protein (MBP) level; after treatment with IVMP, the intrathecal IgM production decreased to low or normal values (Table 1). The reduction in intrathecal IgM and not in intrathecal IgG correlated with the reduction in CSF MBP, indicating that the effect of IVMP treatment is accompanied by a correlative decrease in myelin breakdown and a decrease in IgM production.

In conclusion, CSF IgM and not IgG might be a candidate for establishing disease activity and for monitoring treatment effects.

Cellular immunity

Number of CSF mononuclear cells

Frequin et al. [1] studied the effect of high-dose IVMP treatment in 101 deteriorating MS patients: before treatment the mean number of CSF mononuclear cells had increased in comparison with reference values (6 cells versus 1 cell/μl) (Table 1), while after treatment the mean cell number had decreased by 50%. There was a significant reduction in cell number in RR patients but not in CP patients, and no relation could be found between number of cells and clinical scores (EDSS). More than 2 years after treatment the mean cell number was still reduced (Frequin et al. [2]) (Table 2).

Durelli et al. [16] treated 9 MS patients with high-dose IVMP: after 15 days of treatment the mean mononuclear cell number had reduced from 5.5 to 3.5 cells/μl, and after continuation with low oral MP doses for 25 days the cell number increased to 4.6 cells/μl. Baumhefner et al. [12] could not demonstrate a significant decrease in cell number in 7 low-dose IVMP treated MS patients, and Huber et al. [17] did not find a reduction in CSF cell number in 8 MS patients treated with IV natural IFN-ß.

With respect to immunosuppressants, Uitdehaag et al. [18] compared the short term and long term effects of an intensive immunosuppressive treatment with oral cyclophosphamide; the 26 CP MS patients had clearly increased mean CSF cell numbers before treatment (8 cells/μl). Just after treatment, the cell number had strongly decreased and even more than 10 years later the cell number was still reduced (3 cells/μl), while none of these patients had received any other subsequent treatment. Salmaggi et al. [19] treated 15 CP MS patients for 9 days with high-dose IV cyclophosphamide followed by bi-monthly IV cyclophosphamide; after 1 year of treatment the mean CSF cell number was still reduced from 5 to 2 cells/μl.

In conclusion, intensive IVMP and cyclophosphamide treatment have a short and long term reducing effect on CSF cell number, while IFN-ß and low-dose IVMP treatment have no effect.

Lymphocytes subsets

Several reports have been published on the effect of immunotherapeutic regimens in MS patients on CSF lymphocytes subsets. Dufour et al. [20] treated 9 RR MS patients with high-dose IVMP: there was a non-significant decrease in percentage CSF CD4$^+$ cells, and a significant increase in CD8$^+$ cells (Table 3). The increase in percentage CD8$^+$ cells in CSF was mainly caused by an increase in CD8$^+$ CD28$^+$ cells (T-cytotoxic); a relative decrease in CD8$^+$ CD28$^+$ cells (T-suppressor/effector) was also found.

The decrease in CD4$^+$ was mainly due to a reduction in CD4$^+$ CD45RA$^-$ and not in CD4$^+$ CD45RA$^+$ cells (suppressor-inducer). Durelli et al. [16] studied the effect of high-dose IVMP treatment in 9 RR patients: the percentage CD8$^+$ cells increased significantly after treatment, especially the CD8$^+$ high CD11b$^+$ low type (T-suppressor/effector), and not the CD8$^+$ low CD11b$^+$ high type (T-cytotoxic). This in-

Table 3. CSF T-cell percentages before and at the end of the treatment (high-dose IVMP)

Subsets	Day 0	Day 8
$CD4^+$	71.1 ± 20.6°	56.2 ± 13.8
$CD8^+$	15.8 ± 6.3°	*24.4 ± 6.7
$CD4^+CD45RA^+$	1.9 ± 0.6§	1.6 ± 1.1
$CD4^+CD45RA^-$	68.1 ± 9.6§	*54.7 ± 15.0
$CD8^+CD28^+$	8.9 ± 3.9§	*20.8 ± 7.1
$CD8^+CD28^-$	8.4 ± 6.2§	4.4 ± 1.0

* = denote significant ($p<0.05$) differences between values at day 0 and at day 8; ° = analysis was performed by Wilcoxon-Rank test; § = analysis was performed by Mann Whitney U-test

crease was mainly present in the improving patients; $CD4^+$ cell percentage had decreased. It is obvious that in the study by Dufour the increase in CSF $CD8^+$ concerns the T-cytotoxic $CD8^+$ cell, whereas in the study by Durelli it regards the T-suppressor/effector $CD8^+$ cell. The patients in the study by Durelli et al. [16] were clinically active RR patients, whereas the patients in the study by Dufour et al. [20] had been stable for 3 months before treatment.

Frequin et al. [21] studied the effect of high-dose IVMP in 67 deteriorating MS patients; both in RR and CP patients there was a significant increase in CSF $CD8^+$ cell percentage, a non-significant decrease in $CD4^+$ cell percentage, and a significant decrease in $CD4^+/CD8^+$ ratio after treatment. Among immunosuppressants, Uitdehaag et al. [18] treated 11 CP MS patients with high-dose, orally administered cyclophosphamide; they demonstrated a significant increase in percentage CSF $CD8^+$ cells and a significant decrease in percentage $CD4^+$ cells and $CD4^+/CD8^+$ ratio. Salmaggi et al. [19] reported on a significant increase in percentage CSF $CD8^+$ cells in 15 CP MS patients treated with IV cyclophosphamide for 12 months. Brinkman et al. [22] demonstrated that MS patients treated with high-dose cyclophosphamide had a much lower CSF $CD4^+/CD8^+$ ratio than untreated MS patients.

In contrast to these findings, Polman et al. [23] could not find a lowering effect on CSF $CD8^+$ cells; however, it should be kept in mind that conflicting results in lymphocyte subsets may be caused by the fact that different methods are used (staining by fluorescent or enzyme-linked antibodies, type of antibody used, enumeration of cells by microscopic evaluation or by flow cytometry).

In conclusion, immunotherapy in MS has an increasing effect on percentages CSF $CD8^+$ cells, and a lowering effect on the $CD4^+/CD8^+$ ratio.

Blood CSF barrier

Ratio CSF/serum albumin (Q albumin) is a valid parameter to determine blood-CSF barrier disturbances. It is known that Q albumin values are regularly increased in MS patients, thus forming an indicator of an alteration in the integrity of the blood-CSF barrier. Warren et al. [6] could not demonstrate any change in Q albumin values in 4 treatment groups of active RR MS patients (placebo, IV ACTH, high and mega-dose IVMP); neither did Frequin et al. [1] find a significant reduction in Q albumin in 101 deteriorating MS patients after treatment with high-dose IVMP (Table 1). The same negative result was shown in a study of 26 MS patients treated with IVMP (Anderson et al. [3]). With respect to immunosuppressants, Salmaggi et al. [19] treated 15 CP MS patients for 12 months with IV cyclophosphamide; there was no change in mean Q albumin values 6 months and 12 months after this treatment. Hommes et al. [9] treated 41 CP patients for 20 days with high-dose IV cyclophosphamide and 27 CP patients with chronic low-dose IV cyclophosphamide for 6 months; after treatment there was no significant change in mean Q albumin values.

In conclusion, ratios of CSF/serum albumin values are not influenced by immunotherapeutic approaches.

MBP and anti-MBP antibody

Several Authors have reported on the appearance of CSF MBP levels in patients with MS. Elevated levels of CSF MBP have been found in MS patients with active disease and seem to correspond with disease activity; especially in relapsing MS patients both severity of demyelination and clinical score on EDSS scale were correlated with CSF MBP levels (Frequin et al. [1], Wajgt et al. [10]). Therefore, CSF MBP seems to be a valid parameter for establishing clinical activity in MS and for evaluating treatment effects. Frequin et al. [1] measured MBP levels in CSF in 101 deteriorating MS patients before treatment; MBP was significantly increased in RR patients more than in CP patients. After high-dose IVMP (10 days) the mean increased MBP level had reduced to the reference value (Table 1). 38 patients were also included in a longitudinal prospective study (Frequin et al. [2]); 2.6 years after IVMP treatment the mean CSF MBP level was still reduced from 2.1 μg/L to 0.4 μg/L (Table 2). Barkhof et al. [24] demonstrated in 16 deteriorating MS patients that the number of Gadolinium (Gd-DTPA) enhancing lesions and the CSF MBP level correlated significantly.

After high-dose IVMP there was a significant correlation between decrease in CSF MBP levels, and both decrease in EDSS-scores and in the number of Gd-DTPA enhancing lesions. This may indicate that a reduction in inflammation is accompanied by a decrease in myelin breakdown and clinical recovery. Warren et al. [6] compared the effects of 4 treatment regimens (placebo, low-dose IV ACTH, high-dose IVMP and mega-dose IVMP) on CSF MBP; in the placebo as well as in the ACTH group, CSF MBP levels remained unchanged, while both IVMP treatments resulted in a strong reduction in CSF MBP. The same effects were observed when determining free and total anti-MBP in CSF; increased CSF free and total MBP were significantly reduced only after treatment with IVMP. Also Wajgt et al. [10] could demonstrate in 50 active MS patients that high-dose parenteral prednisone caused a significant reduction in CSF anti-MBP; however, cyclophosphamide in combination with moderate-dose prednisone therapy did not influence anti-MBP levels in the CSF. Lamers et al. [25] studied the short-term effects of a 10 days treatment with cyclophosphamide and prednisone on 11 CP MS patients with a deteriorating form of the disease; before treatment the mean CSF MBP level was clearly increased, after treatment the median value was not significantly different from controls.

Whitaker et al. [26] studied the possible predictive value of CSF MBP measurements; a total of 61 RR and CP patients were treated with IVMP followed by oral prednisone. The results of their study demonstrated that the clinical response of MS patients to IVMP treatment was significantly better in those patients with elevated levels of MBP-like material in CSF. They concluded that CSF MBP-like material provides predictive values as to the outcome of treatment.

In conclusion, CSF MBP levels are regularly increased in active MS patients. Immunotherapy mostly leads to normalization of MBP levels over a relatively long period; CSF MBP level has a predictive value for the outcome of treatment, and the reduction in CSF MBP after treatment correlates with reduction in Gd-DTPA enhancing lesions and clinical scores.

References

1. Frequin STFM, Barkhof F, Lamers KJB, Hommes OR, Borm GF (1992) CSF myelin basic protein, IgG and IgM levels in 101 MS patients before and after treatment with high-dose intravenous methylprednisolone. Acta Neurol Scand 86:291-297
2. Frequin STFM, Lamers KJB, Barkhof F, Brom GF, Hommes OR (1994) Follow-up study of MS patients treated with high-dose intravenous methylprednisolone. Acta Neurol Scand 90:105-110
3. Anderson TJ, MacG Donaldson I, Sheat M, George PM (1990) Methylprednisolone in multiple sclerosis exacerbation: changes in CSF parameters. Aus NZ J Med 20:794-797
4. Wender M, Tokarz E, Michalowska G, Wajgt A (1986) Therapeutic trials of multiple sclerosis and intrathecal IgG production. Ital J Neurol Sci 7:205-208
5. Durelli L, Cocito D, Riccio A, Barile C, Bergamasco B, Baggio GF, Perla F, Delsedime M, Gusmaroli G, Berganini L (1986) High-dose intravenous methylprednisolone in the treatment of multiple sclerosis: clinical-immunologic correlations. Neurol 36:238-243
6. Warren KG, Catz I, Jeffrey VM, Carroll DJ (1986) Effect of methylprednisolone on CSF IgG parameters, myelin basic protein and anti-myelin basic protein in multiple sclerosis exacerbations. Can J Neurol Sci 13:25-30
7. Caputo D, Zaffaroni M, Ghezzi A, Cazzullo CL (1987) Azathioprine reduces intrathecal IgG synthesis in multiple sclerosis. Acta Neurol Scand 75:84-86
8. Staugaitis SM, Shapshak P, Myers LW, Ellison GW, Tourtellotte WW, Lee M (1985) Azathioprine and steroids are not more effective in decreasing multiple sclerosis intra-blood-brain-barrier IgG synthesis than steroids alone. Ann Neurol 18:356-357
9. Hommes OR, Lamers KJB, Geel van WJA (1985) Intrathecal IgG synthesis and IgG index after intensive and chronic immunosuppressive treatment of multiple sclerosis. Ann N Y Acad Sci 436:410-417
10. Wajgt A, Górny M, Szcechowski L, Grzybowski G, Ochudlo S (1989) Effect of immunosuppressive therapy on humoral immune response in multiple sclerosis. Acta Med Pol 30(3-4):121-128

11. Panitch HS, Francis GS, Hooper CJ, Merigan TC, Johnson KP (1985) Serial immunological studies in multiple sclerosis patients treated systemically with human alpha interferon. Ann Neurol 18:434-438
12. Baumhefner RW, Tourtellotte WW, Syndulko K, Staugaitis A, Shapshak P (1989) Multiple sclerosis intra-blood-brain-barrier IgG synthesis: effect of pulse intravenous and intrathecal corticosteroids. Ital J Neurol Sci 10:19-32
13. Confavreux C, Chapuis-Cellier C, Arnaud P, Robert O, Aimard G, Devic M (1986) Oligoclonal "fingerprint" of CSF IgG in multiple sclerosis patients is not modified following intrathecal administration of natural beta-interferon. J Neurol Neurosurg Psych 49:1308-1312
14. Munster van ETL (1991) Thesis, pp 53-71
15. Sharief MK, Thompson EJ (1991) Intrathecal immunoglobulin M synthesis in multiple sclerosis. Brain 114:181-195
16. Durelli L, Poccardi G, Cavallo R (1991) $CD8^+$ high $CD11b^+$ low T cells (T suppressor-effectors) in multiple sclerosis cerebrospinal fluid are increased during high dose corticosteroid treatment. J Neuroimmunol 31:221-228
17. Huber M, Bamborschke S, Assheuer J, Heiß WD (1988) Intravenous natural beta interferon treatment of chronic exacerbating-remitting multiple sclerosis: clinical response and MRI/CSF findings. J Neurol 235:171-173
18. Uitdehaag BMJ, Nillesen WM, Hommes OR (1989) Long-lasting effects of cyclophosphamide on lymphocytes in peripheral blood and spinal fluid. Acta Neurol Scand 79:12-17
19. Salmaggi A, Milanese C, Eoli M, Mantia la L, Nespolo A, Dufour A (1994) Immunological monitoring and clinical evaluation in cyclophosphamide-treated progressive multiple sclerosis patients. Int J Neurosci 76:305-312
20. Dufour A, Salmaggi A, Mantia la L, Eoli M, Nespolo A, Milanese C (1994) High-dose methylprednisolone treatment-induced changes in immunological parameters in progressive MS patients. Int J Neurosci 75:119-128
21. Frequin STFM, Lamers KJB, Barkhof F, Jongen PJH, Hommer OR (1993) T-cell subsets in the cerebrospinal fluid and peripheral blood of multiple sclerosis patients treated with high-dose intravenous methylprednisolone. Acta Neurol Scand 88:80-86
22. Brinkman CJJ, Nillesen WM, Hommes OR (1984) The effect of cyclophosphamide on T lymphocytes and T lymphocyte subsets in patients with chronic progressive multiple sclerosis. Acta Neurol Scand 69:90-96
23. Polman CH, Groot de CJA, Koetsier JC, Sminia T, Veerman AJP (1987) Cerebrospinal fluid cells in multiple sclerosis and other neurological diseases: an immunocytochemical study. J Neurol 234:19-22
24. Barkhof F, Frequin STFM, Hommes OR, Lamers KJB, Scheltens P, Geel van WJA, Valk J (1992) A correlative triad of gadolinium-DTPA MRI, EDSS, and CSF-MBP in relapsing multiple sclerosis patients treated with high-dose intravenous methylprednisolone. Neurol 42:63-67
25. Lamers KJB, Uitdehaag BMJ, Hommes OR, Doesburg W, Wevers RA, Geel van WJA (1988) The short-term effect of an immunosuppressive treatment on CSF myelin basic protein in chronic progressive multiple sclerosis. J Neurol Neurosurg Psych 51:1334-1337
26. Whitaker JN, Layton BA, Herman PK, Kachelhofer RD, Burgard S, Bartolucci AA (1993) Correlation of myelin basic protein-like material in cerebrospinal fluid of multiple sclerosis patients with their response to glucocorticoid treatment. Ann Neurol 33:10-17

Magnetic resonance imaging, proton magnetic resonance spectroscopy and cerebrospinal fluid abnormalities in multiple sclerosis

I.L. Simone, C. Tortorella, P. Giannini, M. Trojano and P. Livrea

The impact of magnetic resonance imaging (MRI)

Magnetic resonance imaging (MRI) is acknowledged as the most sensitive tool for the diagnosis of multiple sclerosis (MS). This technique is not only the procedure of choice in demonstrating plaque dissemination, but it could also allow a prospective evaluation of evolution of the disease processes. Acute and chronic demyelinating plaques appear as areas of increased signal intensity on proton density and T_2-weighted images [1]. In addition, the use of gadopentate dimeglumine (Gd-DTPA), a paramagnetic contrast agent sensitive to changes in blood-brain barrier (BBB) permeability, allows the identification of new or acute lesions in T_1-weighted images. Serial Gd-MRI studies have provided further information regarding the pathogenesis and behaviour of MS demyelinating areas, suggesting that the disease activity may be active even in clinically stable phase [2, 3]. Gd-DTPA enhancement has been advocated as a means of monitoring disease activity since enhanced areas represent an early, and perhaps even the initial, event in new MS lesion development [2, 4]. Correlations between MRI and histopathological findings in experimental chronic encephalomyelitis showed that Gd-enhancement was associated with increased endothelial vesicular transport as a mechanism of BBB breakdown [5]. Furthermore, the level of contrast enhancement was related to the degree of macrophage infiltration more than to perivascular lymphocyte reaction, suggesting that lymphocytes may play a lesser role in BBB breakdown [6].

Histopathology of MS lesions

The earliest detectable event in the development of active MS lesions is a breakdown of the BBB [7], which is associated with acute vasculitis characterized by inflammatory infiltration of the vessel walls. Evidences of ongoing demyelination are myelin fragments and myelin basic protein (MBP) inside the macrophages [8]. In acute early MS disease, oligodendrocytes can be partially preserved within the lesions and a remyelination process is readily detected [9]. In contrast, lesions that develop later in the MS chronic stage are characterized by a very selective destruction

Institute of Clinical Neurology, University of Bari, Policlinico, Piazza Giulio Cesare, 70124 Bari, Italy of both myelin and oligodendrocytes. These data suggest that additional immunological mechanisms, specifically against oligodendrocytes, become pathogenetic with chronicity of the disease.

Axonal injury or destruction is present to a variable degree in all MS lesions [10]. In late chronic MS this process is a common feature, reflected by an increase in the size of the extracellular space, which gives rise to focal or diffuse white matter lesions on MRI. Finally, a progressive and extensive proliferation of astrocytes results in a network which replaces the normal cerebral structures [11]. These data suggest that MS lesions are heterogeneous and change over time: 1) some lesions may be purely edematous, with very little demyelination; 2) some lesions may demyelinate and subsequently remyelinate; 3) other lesions are demyelinated without remyelination.

In spite of the high sensitivity of MR imaging, either alone or with use of Gd-DTPA, this technique provides limited information regarding the pathological heterogeneity of MS lesions. An increase of BBB permeability associated with an inflammatory acute reaction can be well defined by MRI Gd-enhancement, but other processes occurring inside MS lesions, such as demyelination, axonal loss and gliosis cannot be identified. In a recent report, Filippi et al. [12], using conventional T_2-weighted sequences, found a correlation between clinical disability and the number of new or enlarging MRI lesions. In general, however, longitudinal studies showed a poor association between clinical data and MRI findings [3], confirming the limited capacity of MRI in defining the pathological substrate of MS lesions.

^{1}H-MR spectroscopy in brain

Among recent MR techniques, proton MR spectroscopy (^{1}H-MRS) *in vivo* in brain has been proposed as being the most sensitive to characterization of MS lesions, since it provides information on the amount and the quality of pathological changes within the demyelinating plaques over time [13]. This technique makes possible the identification of many metabolites, such as N-acetylaspartate (NAA), creatine-creatinine, choline-containing compounds, inositol, glutammate and glutamine. Changes in these metabolites have been found in brain tumors [14], stroke [15] and MS [13].

^{1}H-MRS also provides the identification of other biochemical compounds as the lactate and mobile lipids, which are detected exclusively in some pathological conditions, depending on specific histopathological processes. In many studies, proton spectra have been acquired from a localized volume of interest (VOI) corresponding to a single MS lesion, approximately 3-8 cm^3 [16, 17], or in some cases even smaller, 1 cm^3 [18]. Other studies reported that the application of ^{1}H-MRS imaging (MRSI) offers a marked improvement over single-voxel spectroscopy since it provides images of regional distribution of metabolites from a large number of small VOI (nominal single voxel size 1 cm^3) [19].

Several reports show that ^{1}H-MRS *in vivo* may be able to distinguish brain acute or active from chronic plaques in MS patients.

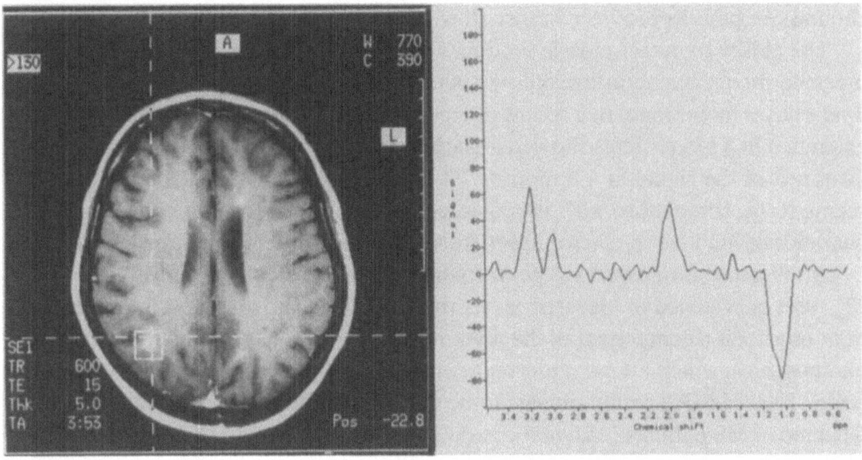

a b

Fig. 1a, b. Proton MR spectrum from localized outlined in axial T_1-weighted MRI Gd-enhanced MS plaque. Resonance of choline at 3.2 ppm, creatine at 3 ppm, NAA at 2 ppm. Note the lactate signal at 1.3 ppm

Metabolic markers of acute inflammatory reaction

An increase in lactate signal may be detected exclusively inside acute MS lesions [19, 20]. Correlation between pathological findings in biopsy specimens and MRSI data confirmed that increased tissue levels of lactate were associated with acute inflammatory reaction in active plaques [19]. Macrophages infiltrating acute plaques seem to be a main source of lactate detected by ^{1}H-MRS investigation [21]. Our preliminary ^{1}H-MRS study, performed on MS lesions with different MRI activity, emphasizes these data showing that lactate signal was more frequent in Gd-enhancing plaques than in Gd-unenhancing plaques [22]. A typical metabolic pattern of an active MS plaque is shown in Fig. 1. Longitudinal studies reported a temporal profile of lactate changes characterized by a gradual disappearance of the signal over 2-3 months [19].

Metabolic markers of acute demyelination

Choline-containing compounds are normal constituents of myelin sheaths and cellular membranes. An increase in the relative resonance intensities of choline has been detected in acute MS plaques [17]. This has been interpreted as evidence of demyelination as well as of increased membrane turnover associated to inflammatory or glial cell reaction [23].

Serial ^{1}H-MRS studies showed an early increase of choline in acute plaques [17, 19] suggesting that demyelination processes occur early in acute MS lesions during BBB breakdown and inflammatory phases. Our recent observations confirmed a significant association between high levels of choline and MRI Gd-enhancing plaques (Simone IL et al., unpublished observations). In follow-up evaluations, choline compounds recovered more slowly than lactate and myo-inositol [20].

The ability to detect mobile lipid resonance in MS lesions is another significant mean to monitor myelin breakdown, since the myelin sheath is constituted of a lipid-bilayer membrane. In a recent prospective study [24] all Gd-enhancing plaques examined in 8 MS patients showed a marked increase of lipid, with a gradual disappearance of the signal in 4-8 months. The time course of ^{1}H-MRS lipid detection seems to be compatible with the histologically determined time course of disappearance of lipid laden macrophages from areas of acute myelin destruction [25].

An increase of inositol has been found within acute enhancing MS lesions [26, 27], with persistence of elevated levels throughout a follow-up of 4-8 months [24]. Myo-inositol is a component of the myo-inositol-containing phospholipids, the phosphoinositides, which are an important constituent of the cellular membranes of the brain. This ^{1}H-MRS result appears to be consistent with high CSF levels of inositol detected in MS patients (enzymatic assay) only during the exacerbation phase [28].

Metabolic markers of axonal degeneration

Several ^{1}H-MRS studies reported a reduction of NAA in MS lesions [17, 19, 24, 26]. The function of NAA is unknown; it is found almost exclusively in neurons and in their processes, and most researchers agree that this metabolite is a a marker of neuronal viability [29]. Chronic demyelinating MS lesions are characterized by a marked irreversible decrease of NAA attributable to an axonal loss [26]. A typical metabolic pattern of inactive chronic MS plaque is shown in Fig. 2. The negative correlation between lower levels of NAA and higher EDSS score [16, 20] suggests that the axonal injury could be an important factor in the development of persistent disability in MS. Some ^{1}H-MRS studies reported a decrease of NAA also in active

a 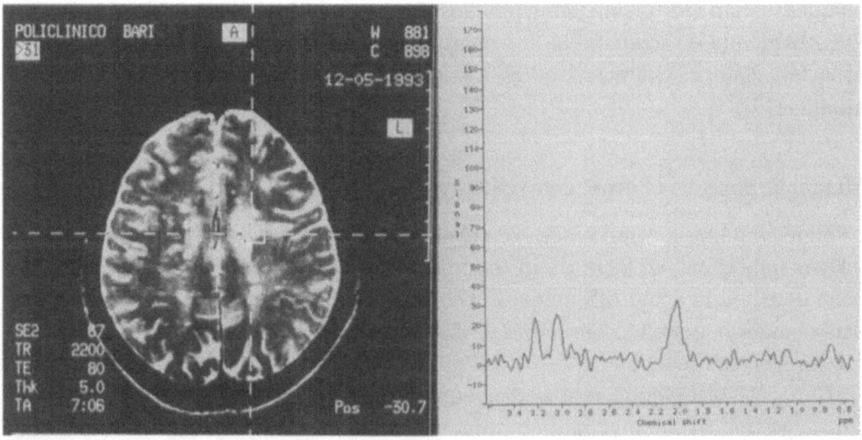 b

Fig. 2a, b. Proton MR spectroscopy from localized VOI outlined in axial T_2-weighted MRI Gd-nonenhanced MS plaques. Resonance of choline at 3.2 ppm, creatine at 3 ppm, NAA at 2 ppm. Note the decrease of NAA signal

plaques, partially reversible over 4-8 months [20, 24]. It is therefore likely that, in the absence of axonal injury, other events may contribute to the reversible changes of NAA in active plaques. These may include a relative reduction of axon number per unit volume depending on edema [7], or a reversible impairment of the function of mitochondria, where NAA is synthesized [30].

^{1}H-MR spectroscopy in CSF

The recent employment of ^{1}H-MRS CSF sample analysis has provided additional information on brain metabolism in normal and pathological conditions [31]. The main metabolites identified in CSF by ^{1}H-MRS are alpha and beta glucose, lactate, acetate, citrate, formate, creatine-creatinine (Fig. 3).

Few ^{1}H-MRS studies have been performed to investigate biochemical changes in CSF of MS patients [31, 32]. In one preliminary report [22], the ^{1}H-MR CSF spectra of MS patients were compared with their *in vivo* brain spectra, and the relationships between metabolic changes and clinical or MRI activity were investigated (Table 1). High CSF lactate levels were found both in MS patients during clinical exacerbation and in MS patients with MRI evidence of Gd-enhancing plaques. In the latter group a high lactate signal was detected by spectroscopy *in vivo* on MRI active plaques. CSF formate and citrate levels were decreased in MS patients inde-

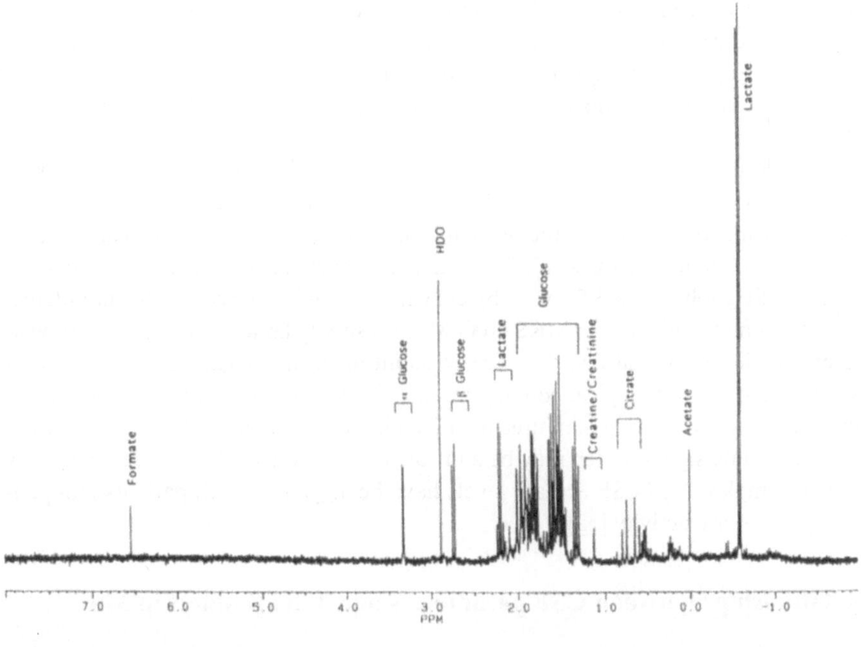

Fig. 3. Proton MR spectrum in CSF sample. Labelled resonance include residual water proton (HDO), alpha-beta glucose, lactate, citrate, formate, creatine-creatinine

Table 1. ^{1}H-MRS metabolites in brain and CSF. Relationships between biochemical changes and clinical and MRI activity in MS

	BRAIN		CSF	
	MRI normal brain	**MS plaques**	**Control samples**	**MS samples**
Lactate	undetectable	↑ in *active* plaques	present	↑ in patients with *clinical or MRI activity*
Choline	present	↑ in *active* plaques	undetectable	undetectable
Creatine	present	stable	present	stable
NAA	present	↓ in *chronic* plaques	absent	absent
Acetate	absent	absent	present	↑ independently *of clinical* or *MRI activity*
Formate	absent	absent	present	↓ independently *of clinical* or *MRI activity*
Lipids	undetectable	↑ in *active* plaques	unidentified signal	unidentified signal
Citrate	undetectable	undetectable	present	↓ independently *of clinical* or *MRI activity*
Myo-inositol	present	↑ in *active* plaques	undetectable	undetectable

↑ = increase
↓ = decrease

pendently of clinical phase or MRI plaque activity; the formate signal is detected exclusively in CSF, whereas there is no distinguishable resonance in the brain spectra. It has been suggested that formate could be an intermediary metabolite of choline and its decrease in CSF may depend on an impairment of the choline pathway in MS [32].

^{1}H-MR spectra, acquired both in normal and pathological CSF samples, showed absence of signals corresponding to choline and NAA resonances. These metabolites are important constituents of the normal brain, and their biochemical changes have been detected in MS plaques [17, 19, 24, 26], as reported above. The low concentrations of CSF choline (1.5-3.5 μM/L, by enzymatic assay) may account for non-detection of this metabolite by ^{1}H-MRS (100μM/L sensitivity limit). NAA is an unexpected metabolite in normal CSF because of its intraneuronal location; the absence of NAA in CSF in pathological conditions, such as MS, could depend on NAA enzymatic hydrolysis occurring in injured axons, and, probably at a greater extent, in CSF.

The acetate signal, which may be a product of NAA hydrolysis, is detected only in CSF samples; high CSF acetate levels have been found in MS patients independently of disease activity [32].

Relationships between CSF parameters and MRI findings in MS

CSF analysis yields typical abnormalities in MS which reflects an immunological impairment. The most frequent CSF alteration is an abnormal B-cell response within

Table 2. Relationships between MRI findings and CSF parameters

	Total MRI lesion area (Baumhefner [40])	**Volume of MRI periventricular lesions** (Kappos [41])	**Number of MRI non-periventricular lesions** (Baum [42])	**Number of Gd-DTPA lesions** (Barkhof [43])	
				before MP iv	after MP iv
Mononuclear cells				n.s.	n.s.
Q albumin			$p < 0.05$	n.s.	n.s.
Oligoclonal bands	n.s.	$p = 0.05$		n.s.	$p = 0.05$
IgG synthesis	$p < 0.01$	$p = 0.05$	$p < 0.05$	n.s.	n.s.
CSF-IgG		$p = 0.05$	$p < 0.05$		
CSF-IgA			$p < 0.05$		
CSF-MBP				$p = 0.004$	$p = 0.001$

CNS, as shown by elevated CSF IgG levels and oligoclonal IgG fractions [33]. The antibody specificities of the intrathecally synthesized IgG are not well defined; only a minor proportion is constituted by IgG antibodies against some neurotropic viruses [34].

Many CSF antibodies have been identified against various proteins of CNS such as myelin basic protein (MBP) [35], proteolipid protein (PLP) [36], myelin oligodendrocyte glycoprotein - (MOG) [37], and myelin associated glycoprotein - (MAG) [38]. The evidence of these auto-antibodies and the presence of myelin protein-like material in CSF [39] have been proposed as a means of monitoring the pathological changes occuring in demyelinating plaques.

Immunological studies on the CSF of MS patients showed an intrathecal synthesis of anti-MOG antibodies in the chronic stage of the disease, characterized by selective oligodendrocyte destruction [36]. A study performed in a large series of MS patients, with different clinical courses and phases of the disease, reported that high CSF-MBP levels were significantly associated with the exacerbation phase, its duration and the clinical disability of the exacerbation [39].

The relationships between MRI findings and CSF abnormalities in MS have been evaluated only in a few studies (review in Table 2).

In a series of 62 MS patients with clinically definite chronic progressive MS, Baumhefner et al. [40], using a technique of quantification of MRI abnormal signal intensity areas, reported that the total cerebral area of MRI lesions correlated significantly with intrathecal IgG synthesis rate, but not with the number of CSF oligoclonal IgG fractions. In a preliminary study, Kappos et al. [41] found a low but significant correlation of CSF IgG levels and IgG index, as well as of CSF oligoclonal fractions and MRI periventricular lesions, but not with total cerebral lesion volume. The close proximity between periventricular lesions and CSF space could account for these data. A correlation between the number of non-periventricular lesions and CSF IgG, IgA levels and also IgG intrathecal synthesis rate has been shown by Baum et al. [42]. Furthermore, in this study CSF albumin concentrations were significantly correlated with both the number of non-periventricular lesions and the extent of periventricular involvement expressed by periventricular score. No correlation was found between MRI morphological findings and the number of mononuclear cells.

Only in a very few studies, the relationship between Gd-DTPA enhancement and CSF abnormalities has been investigated. In a prospective study performed in 16 relapsing-remitting MS patients, Barkof et al. [43] compared the number of Gd-enhancing lesions with CSF parameters and clinical findings before and after high-dose intravenous (i.v.) of methylprednisolone (MP) treatment. The number of Gd-enhancing lesions significantly correlated with CSF-MBP levels, suggesting that inside MS plaques inflammation and blood-brain barrier breakdown are directly associated with demyelination. These results appear to be consistent with ^{1}H-MRS data, which showed an early demyelination in the development of MS lesions, as above reported [17, 19, 24]. No correlation between Gd-enhancing lesions and oligoclonal bands was found. After i.v. MP, the decrease in the number of Gd-enhancing lesions was significantly associated with the decrease in CSF-MBP levels and EDSS score. No changes in numbers of oligoclonal bands, Q albumin and mononuclear cells were found after MP therapy.

In conclusion, the finding of relationships between CSF parameters, total lesion load on T_2-weighted MRI, Gd-DTPA enhanced lesions, ^{1}H-MRS features and clinical data indicates that these variables may provide a complementary view of the underlying disease activity in MS patients. Each set of these measurement may well become important in evaluating the progression of the disease and in monitoring the efficacy of treatment on inflammatory, demyelination, axonal loss and gliosis processes.

References

1. Paty DW, Oger JJF, Kastrukoff LF, Hashimoto SA, Hooge JP, Eisen AA, Eisen KA, Purves SJ, Low MD, Brandejs V, Robertson WD, Li DKB (1988) MRI in the diagnosis of MS: a prospective study with comparison of clinical evaluation, evoked potentials, oligoclonal banding, and CT. Neurology 38: 180-185
2. Kermode AG, Tofts PS, Thompson AJ, MacManus DG, Rudge P, Kendall BE, Kingsley DPE, Moseley IF, du Boulay EPGH, McDonald WI (1990) Heterogeneity of blood-brain barrier changes in multiple sclerosis: an MRI study with gadolinium-DTPA enhancement. Neurology 40: 229-235
3. Harris JO, Frank JA, Patronas N, McFarlin DE, McFarland HF (1991) Serial gadolinium-enhanced magnetic resonance imaging scans in patients with early, relapsing-remitting multiple sclerosis: implications for clinical trials and natural history. Ann Neurol 29: 548-555
4. Grossmann RI, Braffman BH, Brorson JR, Goldberg HI, Silberberg DH, Gonzales-Scarano F (1988) Multiple sclerosis: serial study of gadolinium-enhanced MR imaging. Radiology 169: 117-122
5. Hawkins CP, Mackenzie F, Tofts P, du Boulay EPGH, McDonald WI (1991) Patterns of blood-brain breakdown in inflammatory demyelination. Brain 114: 801-810
6. Nesbit GM, Forbes GS, Scheithauer BW, Okazaki H, Rodriguez M (1991) Multiple sclerosis: histopathologic and MR and/or CT correlation in 37 cases at biopsy and three cases at autopsy. Radiology 180: 467-474

7. McDonald WI, Miller DH, Barnes D (1992) The pathological evaluation of multiple sclerosis [review]. Neuropathol Appl Neurobiol 18: 319-334
8. Prineas JW, Kwon EE, Cho E-S, Sharer LR (1984) Continual break-down and regeneration of myelin progressive multiple sclerosis plaques. Ann NY Acad Sci 436: 11-32
9. Prineas JW, Barnard RO, Kwon EE, Sharer LR, Cho E-S (1993) Multiple sclerosis: remyelination of nascent lesions. Ann Neurol 33: 137-151
10. Roizin L, Haymaker W, D'Amelio F (1982) Disease states involving the white matter of the central nervous system. In: Haymaker W, Adams RD (eds) Histology and histopathology of the nervous system. Springfield, III, Thomas, pp 1276-1324
11. Duchen LW (1984) General pathology of neurons and neuroglia. In: Adams JH, Corsellis JAN, Duchen LW (eds) Greenfield's Neuropathology. London, Edward Arnold, pp 1-52
12. Filippi M, Paty DW, Kappos L, Barkhof F, Compston DAS, Thompson AJ, Zhao GJ, Wiles CM, McDonald WI, Miller DH (1995) Correlations between changes in disability and T_2-weighted brain MRI activity in multiple sclerosis. A follow-up study. Neurology 45: 255-260
13. Richard TL (1991) Proton MR spectroscopy in multiple sclerosis: value in establishing diagnosis, monitoring progression, and evaluating therapy. AJR: 1073-1078
14. Kugel H, Heindel W, Ernestus R, Bune J, du Mesnil R, Friedman G (1992) Human brain tumors: spectral patterns detected with localized H-1 MR spectroscopy. Radiology 183: 701-709
15. Henriksen O, Gideon P, Sperling B, Skyhøj T, Jøgensen HS, Arlien-Søborg P (1992) Cerebral lactate production and blood flow in acute stroke. J Magn Reson Imaging 2: 511-517
16. Arnold DL, Matthews PM, Francis G, Antel J (1990) Proton magnetic resonance spectroscopy of human brain *in vivo* in the evaluation of multiple sclerosis: assessment of the load of disease. Magn Reson Med 14: 154-159
17. Matthews PM, Francis GS, Antel J, Arnold DL (1991) Proton magnetic resonance spectroscopy for metabolic characterization of plaques in multiple sclerosis. Neurology 41: 1251-1256
18. Frahm J, Michaelis T, Merboldt KD, Bruhn H, Gyngell ML, Hänicke W (1990) Improvements in localized proton NMR spectroscopy of human brain. Water suppression, short echo times and 1 ml resolution. J Magn Reson 290: 464-473
19. Arnold DL, Matthews PM, Francis GS, O'Connor J, Antel JP (1992) Proton magnetic resonance spectroscopic imaging for metabolic characterization of demyelinating plaques. Ann Neurol 31: 235-241
20. De Stefano N, Matthews PM, Antel JP, Preul M, Francis G, Arnold DL (1995) Chemical pathology of acute demyelinating lesions and its correlation with disability. Ann Neurol 38: 901-909
21. López-Villegas D, Lenkinski RE, Wehrli SL, Ho WZ, Douglas SD (1995) Lactate production by human monocytes/macrophages determined by proton MR spectroscopy. Magn Reson Med 34: 32-38
22. Simone IL, Tortorella C, Liguori M, Giannini P, Picciola E, Carrara D, Federico F, Livrea P (1995) Biochemical changes in cerebrospinal fluid (CSF) and demyelinating area in multiple sclerosis (MS). Evaluation by proton magnetic resonance spectroscopy (^{1}H-MRS). J Neuroimmunol 1: 73
23. Miller BL (1991) Review of chemical issue in ^{1}H-NMR spectroscopy: N-acetyl-L- aspartate, creatine and choline. NMR Biomed 4: 47-52
24. Davie CA, Hawkins CP, Barker GJ, Brennan A, Tofts PS, Miller DH, McDonald WI (1994) Serial proton magnetic resonance spectroscopy in acute multiple sclerosis. Brain 117: 49-58

25. Adams CWM, Poston RN, Buk SJ (1989) Pathology, histochemistry and immunocytochemistry of lesions in acute multiple sclerosis. J Neurol Sci 92: 291-306
26. Bruhn H, Frahm J, Merboldt KD, Hänicke W, Hanefeld F, Christen HJ, Kruse B, Bauer HJ (1992) Multiple sclerosis in children: cerebral metabolic alterations monitored by localized proton magnetic resonance spectroscopy *in vivo*. Ann Neurol 32: 140-150
27. Koopmans RA, Li DKB, Zhu G, Allen PS, Penn A, Paty DW (1993) Magnetic resonance spectroscopy of multiple sclerosis: *in vivo*-detection of myelin breakdown product [letter]. Lancet 341: 631-632
28. García-Buñuel L, García-Buñel VM (1965) Cerebrospinal fluid levels of free myoinositol in some neurological disorders. Neurology 15: 348-350
29. Simmons ML, Frondoza CG, Coyle JT (1991) Immunocytochemical localization of N-acetyl-aspartate with monoclonal antibodies. Neuroscience 45: 37-45
30. Patel TB, Clark JB (1979) Synthesis of N-acetyl-L-aspartate by rat brain mitochondria and its involvement in mitochondrial/cytosolic carbon transport. Biochem J 184: 539-546
31. Koschorek F, Offermann W, Stelten J, Braunsdorf WE, Steller U, Gremmel H, Leibfritz D (1993) High-resolution ^{1}H-NMR spectroscopy of cerebrospinal fluid in spinal diseases. Neurosurg Rev 16: 307-315
32. Lynch J, Peeling J, Auty A, Sutherland GR (1993) Nuclear magnetic resonance study of cerebrospinal fluid from patients with multiple sclerosis. Can J Neurol Sci 20: 194-198
33. Livrea P, Trojano M, Simone IL, Zimatore GB, La Montanara G, Leante R (1981) Intrathecal IgG synthesis in multiple sclerosis: comparison between isoelectric focusing and quantitative estimation of cerebrospinal fluid IgG. J Neurol 224: 159-169
34. Norrby E (1978) Viral antibodies and multiple sclerosis. Progr Med Virol 241: 1-39
35. Warren KG, Catz J (1989) Cerebrospinal fluid autoantibodies to myelin basic protein in multiple sclerosis patients. J Neurol Sci 91: 143-151
36. Sun JB (1993) Autoreactive T and B cells in nervous system diseases. Acta Neurol Scand S142: 1-56
37. Xiao G-B, Linington C, Link H (1991) Antibodies to myelin-oligodendrocyte glycoprotein in cerebrospinal fluid from patients with multiple sclerosis and controls. J Neuroimmunol 31: 91-96
38. Baig S, Olsson T, Jiang Y-P, Höjerberg B, Cruz M, Link H (1991) Multiple sclerosis: cells secreting antibodies against myelin-associated glycoprotein are present in cerebrospinal fluid. Scand J Immunol 33: 73-79
39. Warren KG, Catz I (1985) The relationship between levels of cerebrospinal fluid myelin basic protein and IgG measurements in patients with multiple sclerosis. Ann Neurol 17: 475-480
40. Baumhefner RW, Tourtellotte WW, Syndulko K, Waluch V, Ellison GW, Meyers LW, Cohen SN, Osborne M, Shapshak P (1990) Quantitative multiple sclerosis plaque assessment with magnetic resonance imaging. Arch Neurol 47: 19-26
41. Kappos L, Pfeuffer B, Staedt D, Rohrbach E, Heun R, Haubitz I, Keil W (1988) Quantitative magnetic resonance imaging of the multiple sclerosis brain: localization rather than total volume of lesions correlates with cerebrospinal fluid IgG level and visual evoked response abnormalities. Ann Neurol 24: 169
42. Baum K, Nehrig C, Girke W, Bräu H, Schörner W (1990) Multiple sclerosis: relations between MRI and CT findings, cerebrospinal fluid parameters and clinical features. Clin Neurol Neurosurg 92: 49-56
43. Barkhof F, Frequin STFM, Hommes OR, Lamers K, Seheltens P, van Geel WJA, Valk J (1992) A correlative triad of gadolinium-DTPA, and CSF-MBP in relapsing multiple sclerosis patients treated with high-dose intravenous metylprednisolone. Neurology 42: 63-67

Quality assurance and sample handling in cerebrospinal fluid investigation

S. Öhman

Clinical chemistry begins and ends with two decisions: a) to take a sample and b) to act according to the result of the analysis. In between a series of events take place; each step involves a certain risk of making mistakes. The art of avoiding such mistakes is called *Quality Assurance* (QA).

The sample usually consists of a body fluid taken in order to analyse it with respect to one or more components (analytes). Most studies in clinical chemistry deals with blood samples, and obviously such QA systems apply also to those blood samples (whole blood, plasma or serum) taken in multiple sclerosis (MS) patients. Many QA routines developed for blood samples can also be applied to cerebrospinal fluid (CSF).

Knowledge

The first as well as the last steps of QA chain are based on knowledge of how the chemical analysis is linked to the disease; rarely one single chemical analysis is pathognomic. For MS patients the diagnosis is mainly set by the clinical symptoms, but laboratory tests are important aids in this investigation [1, 2]. The most important value of laboratory tests is not their ability to *confirm* MS, but their ability to *exclude* MS in patients where the clinical symptoms are inconclusive. Although the knowledge of the ability of different methods to confirm or exclude a disease is not generally included in QA, it is nevertheless important and should be considered in this context.

For MS, determinations of immunoglobulins (Ig) in CSF are of diagnostic value; most interest is focused on immunoglobulin G (IgG), but determinations of IgA and IgM should be considered as optional methods [2-4]. The results of CSF analysis must always be compared with of the corresponding serum in order to differentiate between intrathecal processes and systemic [2]. Hence, the reliability of the test is dependent on both the CSF and serum methods.

Department of Clinical Chemistry, Faculty of Health Sciences, University of Linköping, S-581 85 Linköping, Sweden.
Present address: ELFINILAB, Po Box 133, S-59070 Ljungsbro, Sweden

Sampling of CSF

The details of the sampling technique are seldom sufficiently described, although no analytical method can compensate for a bad sampling. Sampling is performed by lumbar puncture (LP), and a number of techniques are in use.

A good QA needs a detailed description of the sampling procedure and how it affects the analytical result. This fact is not commonly considered: a MEDLINE search covering the period 1966-1992 yielded no reference on this matter.

An important source of error, especially in the case of IgM, is contamination with serum. Erythrocyte count is commonly used as an indicator of blood contamination but, in a recent study, we found erythrocytes and IgM in CSF to be completely non-correlated [5]. This means that CSF may be contaminated with serum without any increase of erythrocyte count. Probably this error depends on the sampling technique [5].

One often unforeseen source of error in the sampling procedure is that the composition of the CSF in the lumbar sac differs from that of other parts of the central nervous system. It has long been known that the CSF of the ventricles has approximately half of the lumbar concentration of proteins [6]. Furthermore, there is a concentration gradient within the spinal canal, which means that the concentration of the sample varies with the volume of the sample [7]; therefore a defined amount of CSF (preferably 10 mL for adults) should be collected [2]. A schematic view of the CSF flow together with aspects on CSF analysis was recently given by Watson and Scott [8]. The effect of the CSF flow on the reliability of CSF analysis can be illustrated by an occasional patient with subarachnoid bleeding. The CSF in the lumbar sac did not contain any trace of blood; CT scan revealed that the bleeding was situated close to the sagittal vein. Obvioulsy the blood was immediately transported by the CSF to the arachnoid villi and back into the blood, and therefore it never reached the lumbar sac.

There is a small (<5%) but significant number of MS patients where all current tests, e.g. oligoclonal IgG bands [2, 9], in CSF are normal. This fact may be due to the localisation of MS lesions to places where the CSF does not reach or mix with the contents of lumbar sac.

Storage and transport

A well-known source of error is adsorption of proteins to the surface of the tube; therefore CSF should be collected and stored in polypropylene, siliconised glass, or glass tubes [2].

Cells (i.e. erythrocytes and leukocytes) rapidly change their shapes after sampling, and they should preferably be counted within 30 minutes, and never later than 2 hours from sampling. Malignant cells are usually considered as more sturdy than normal cells; therefore CSF samples for determination of such cells frequently are sent separately (e.g. by post) to special laboratories. However, a recent contact with such a laboratory (unpubl) revealed that many samples contained cells of inferior quality, causing problems in the correct diagnosis.

As soon as the CSF or serum sample is taken the temperature decreases from 37 °C to ambient or refrigerator temperature. At low temperatures IgA and especially IgM have a tendency of forming aggregates; these can sedimentate, and after a couple of days the concentration is lower in the top of the tube than in the bottom.

Freezing and thawing induce shear stress to the proteins and may damage them. These measures should therefore be performed rapidly, minimizing the time at temperatures about 0 °C (freezing) and -11 °C (recrystallizing of ice). Before freezing, the sample should be kept at 37 °C for at least 30 minutes in order to dissociate any aggregates, then it should be rapidly frozen to -80 °C preferably in dry ice/ethanol. Storing can then be performed at -20 °C, provided that the freezer never allows the temperature to rise above -15 °C. Thawing should be performed at 37-40 °C (water bath) to assure a rapid temperature increase.

Storing and transport at ambient temperature should be avoided, because of the risk of bacterial growth. Maximum storage time should be 72 hours at 4 °C or 12 hours at 25 °C; if samples should be stored for a longer time, they should be kept in a freezer or transported with dry ice.

Analytical errors

This part of QA is generally called *Quality Control* (QC), and it is since long established within clinical chemistry [10, 11]. QC should be applied to all methods irrespective whether it is quantitative (e.g. CSF-Albumin) or qualitative (e.g. IgG isoelectric focusing). Every run should contain at least one internal control (i.e. a locally or commercially produced sample intended to detect analytical errors). Furthermore, external controls (*Ringversuch*, i.e. samples distributed blindly to several laboratories and evaluated against a reference method or a consensus value) should be run periodically [12].

Cytological methods. Because of the rapid change in cell morphology after LP (*vide supra*), conventional QC programs cannot be used. Sending sets of slide preparations has been proposed [2], but has not been done so far. The lack of a good QC program for cytological methods is not helpful, and efforts should be made to establish such a program.

Qualitative methods. A typical qualitative method used in the diagnosis of MS is determination of oligoclonal IgG bands (see section 4). For a qualitative method there are only two possible results: presence and absence, e.g. of oligoclonal IgG bands. Each run, therefore, needs two internal controls, one with a known oligoclonal pattern (positive control) and one containing only polyclonal IgG (negative control). The method should distinguish between no bands and faint bands, and therefore the positive control should include bands which disappear as soon as the method is not working satisfactorily.

Inhomogenities in the gel and/or the ampholytes may cause artefactual bands to appear. Therefore, the negative control is equally important to include as the posi-

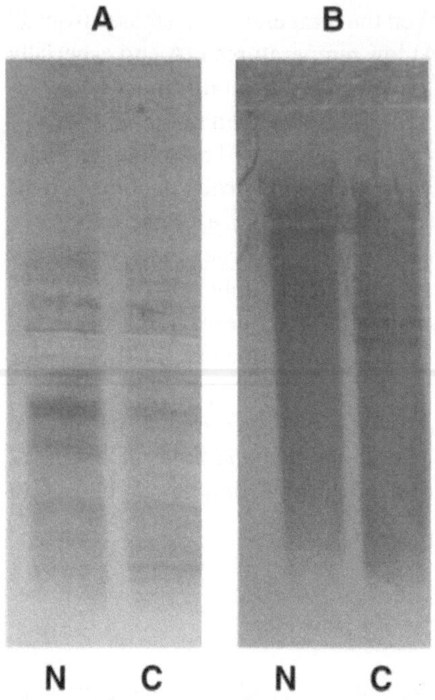

Fig. 1A, B. Internal controls for isoelectric focusing in agarose of oligoclonal IgG. N is a negative (normal) control and C is a positive control (diluted serum) containing oligoclonal bands. In **A** the method of Olsson et al. [13] was used, and in **B** that of Keir et al. [14]. In the former method a number of false bands appeared for the negative control due to the unlinear pH gradient using Ampholines®. In the latter method Pharmalytes® were used yielding a linear pH gradient, and consequently no false bands appeared

tive. Fig. 1 shows isoelectric focusing in agarose gel with Ampholines® (Fig. 1A) and Pharmalytes® (Fig. 1B) run with otherwise similar techniques. As can be seen, the former displays several artefactual bands due to non-homogenous pH gradient, whether the latter displays the polyclonal lanes as a homogenous band [14].

Quantitative methods. There are two kinds of analytical errors: systematic and random errors. Many QC systems deal only with the latter, but both are important. The result of an analytical procedure (V) deviates from the absolute correct value ($V°$) according to the following equation:

$$V = V° + \beta + \varepsilon \qquad (1)$$

where β is the systematic error (*bias*) and ε is the random error. β and ε are algebraic terms, i.e. they can be positive as well as negative numbers. In QC systems both terms should be investigated, although different methods are used. Quantitative methods can have higher or lower *status* according to Fig. 2. An analytical method should always be traceable to a better method in a higher level, and ultimately to the SI units [15]. The latter is the only method without systematic errors, or a "fixed point" to which other methods are calibrated.

Systematic errors. The methods in current use for CSF analysis should conform to the rules of good laboratory practice (GLP). For details see specialised literature on this matter [16]. A GLP method must be calibrated against a reference method

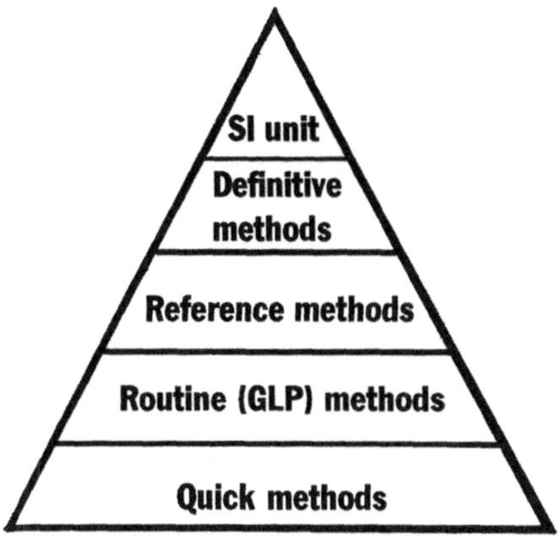

Fig. 2. Hierarchy of analytical methods. *At the top* there are the definitions of the SI units; a little lower in rank are other definitions, e.g. WHO standards which are assigned exact values. Reference methods are not intended to be used routinely; they are accurate and often tedious methods, used in a limited number of laboratories and intended for calibration of other methods. In clinical routine the methods shall conform to the rules of Good Laboratory Practice (GLP), as established by international authorities. *At the bottom* of the hierachy reside methods with considerable inaccurracy and/or imprecision, but which can be performed quickly and/or near the patient. As a general rule, every method shall be traceable to a method with higher rank, and eventually to the SI units at the top

(Fig. 2), which in turn is calibrated against a definitive method. For proteins this means that they should be traceable to WHO standards [15], e.g. CMR 470. The systematic deviation from this standard should be known.

Systematic errors can be determined by multiple determinations of a control sample with a known concentration as determined by a reference method. For an infinite number of determinations the random error, ε in equation (1), for the mean ($\bar{V}\infty$) is zero. Hence, the systematic error can be calculated:

$$\beta = \bar{V}\infty - V^{\circ} \qquad (2)$$

Systematic errors are not merely limited to the analysis procedure, but should be considered within the whole chain of events, i.e. sampling, transport, storage, analysis and data acquisition.

Random errors. Usually ε in equation (1) is considered to be Gaussian distributed. Often this assumption is apparently true, but the possibility of irregularly distributed ε should always be considered. This means that the errors tend to deviate considerably more in positive than in negative direction or *vice versa*. Such a biased distribution is easy to disclose. Usually the results of the control runs are registered in a Shewhart plot [11, 17]; each month (or any other suitable interval) the mean, standard deviation, and coefficient of variation [17] is calculated. If ε is symmetrically distributed, equal numbers of observations should appear below and above a symmetrical range, e.g. mean ± 1SD. This can be tested by a binomial test on $p = 0.5$ (sign test, see Miller and Miller [17] or any other textbook in statistics for details).

The random error is usually expressed as a standard deviation or as a coefficient of variation. Note that many methods have a narrow optimal concentration range; the control samples should be chosen not according to this optimal range, but to the concentrations which are important for the clinical decisions, e.g. about the reference limits.

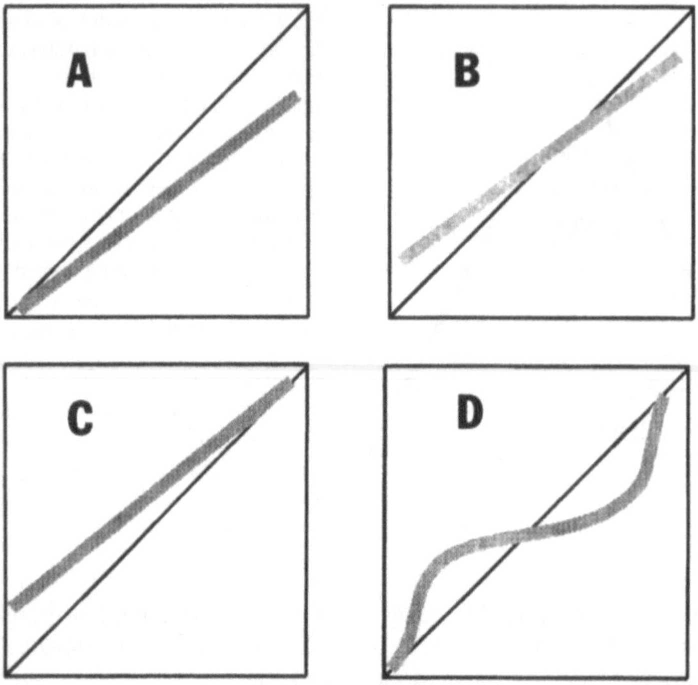

Fig. 3A-D. Different kinds of methodological errors. The "correct" values are indicated on the abscissa and the observed values on the ordinate. The correct values may be hypothetical, determined by a reference method (see Fig. 2), or established by a consensus. The *diagonal lines* indicate the ideal case, where all observed values are absolutely correct; *shaded broad lines* indicate different kinds of analytical errors that may occur. In **A** there is a proportional systematic error (bias), e.g. caused by inappropriate calibration; **B** and **C** show the effect of combined absolute (i.e. independent of the value) and proportional bias; **D** shows a non-linear response

Formulas. In CSF analysis different formulas are often used [18]; they are equations where the results of two or more CSF and serum analyses are combined in order to determine the blood-CSF barrier (e.g. the albumin quotient), or compensate for variations in this barrier (e.g. IgG index). The combined effect of errors in the primary determinations can sometimes be quite complicated. This object has earlier been reviewed for formulas used in CSF analysis [17].

As illustrated in the following examples, evaluation of the analytical error of a formula needs knowledge on systematic as well as the random errors for each included component. Here follows a mathematical explanation of some current cases:

1) The formula includes only transformations and constants

An example of this case is the difference between IgG index [18] and Log IgG index [19] where:

$$\text{Log IgG index} = \text{In (IgG index)} + 1 \qquad (3)$$

In a non-parametric model, these two formulas are equivalent. Neither taking the logarithm nor the addition of the constant can change the rank order of a series of IgG indices. All systematic and random errors are conserved after the transformation, although the absolute magnitude of the bias and the shape of the distribution of the random error is changed.

2) The formula includes the difference between two analytical results

Let the true values of the two methods be V°_1 and V°_2, and the analytical results V_1 and V_2. Then:

$$V_1 = V^{\circ}_1 + \beta_1 + \varepsilon_1 \qquad (4)$$

$$V_2 = V^{\circ}_2 + \beta_2 + \varepsilon_2 \qquad (5)$$

$$V_1 - V_2 = V^{\circ}_1 - V^{\circ}_2 + \beta_1 - \beta_2 + \varepsilon_1 - \varepsilon_2 \qquad (4, 5)$$

If the two methods are similar, the systematic errors (β_1 and β_2) are expected to be similar. If all systematic errors have the same sign (Figs. 2A and 2C) the bias tends to diminish after subtraction. However, if the errors are like those in Figs. 2B or 2D, the subtraction can increase the bias.

The random error always increases after subtraction. Assuming ε_1 and ε_2 to be Gaussian distributed with the standard deviations s_1 and s_2, the combined standard deviation ($s_{1,2}$) for $\varepsilon_1 - \varepsilon_2$ is:

$$s_{1,2} = \sqrt{s_1^2 + s_2^2} \qquad (6)$$

The effect on the coefficient of variation (CV) is more pronounced; CV (here shown as a dimensionless variable, but frequently expressed as a percentage) is defined as the standard deviation divided with the mean value [17]:

$$CV = \frac{s}{V} \qquad (7)$$

Thus, the CV for the difference between $\bar{V}_1$ and $\bar{V}_2$ is:

$$CV_{1,2} = \frac{s_{1.2}}{\bar{V}_1 - \bar{V}_2} \qquad (8)$$

Obviously, if the two terms are of the same magnitude, $CV_{1.2}$ can be considerably high. This is the reason why formulas based on differences (e.g. Tourtellotte's formula), although based on physiologically sound theory, often yield unreliable clinical results, as reviewed by Lefvert and Link [18].

3) The formula includes the quotient between two analytical results

When using CSF/serum quotients and formulas derived from such quotients, it is often assumed that the analytical errors are eliminated, or at least diminished, especially if the two determinations were performed with the same technique and in the same run [12]. However, quotients eliminate only one kind of error, *viz.* the calibration error (Fig. 2A). This kind of systematic error is proportional to V. There are other kinds of errors (matrix effects, non-linearity and non-specific interactions, Figs. 2B, 2D) that are not proportional to V and therefore must be considered.

Combining equations (3) and (4), using division yields:

$$Q(V) = \frac{V_1}{V_2} = \frac{V_1{}^{\circ}+\beta_1+\varepsilon_1}{V_2{}^{\circ}+\beta_2+\varepsilon_2} \qquad (9)$$

Obviously, the error of the quotient depends on the magnitude of all terms in the equation. If we assume that the random error in the denominator (ε_2) is zero, then this error of Q(V) depends only on the numerator. If ε_1 is Gaussian distributed, also the random error of Q(V) will have a Gaussian shape. However, in the opposite situation, i.e. $\varepsilon_1 = 0$, the random error of Q(V) is skewed.

A proportional systematic error (Fig. 2A) is eliminated in Q(V), whereas other kinds of systematic errors (Figs. 2B-D) may enhance or diminish the systematic error of Q(V) depending on the magnitude of the terms.

4) The formula includes more than two analytical results

The above mentioned rules can be applied to more complicated formulas. However, the number of unpredictable terms increases and an empirical approach may be as good as a mathematical one. The following non-parametric design has the advantage of assuming nothing about the distribution of errors.

Run one control sample of CSF and one of serum as if they were real (paired) samples, and calculate the formula. When a sufficient number (at least 120) of determinations is run, calculate the median and the central 95% fractile. The latter is calculated non-parametrically. Rank the results in increasing order; determine how many results that fall outside the range, i.e. 2.5% of the observations in each tail: for 120 determinations there should be 6 values outside the range, i.e. 3 in each tail.

Compare the median value with the assigned value for the control; the difference is assumed to be the systematic error of the formula. Compare the 95% limits with the maximum allowable error that could be accepted for clinical reasons. If this range is too large, the methods have to be improved.

Conclusion

A satisfactory QA system is essential in all analytical methods, including CSF analysis. Quality must be maintained through the whole series of events, from the decision to take the sample to the evaluation of the result. Although the analytical quality is very important, a professional approach in all steps of the procedure is mandatory.

References

1. Poser CM, Paty DW, Scheinberg L, McDonald I, Davis FA, Ebers GC, Johnson KP, Sibley WA, Silberberg DH, Tourtellotte WW (1983) New diagnostic criteria for multiple sclerosis: Guidelines for research protocols. Ann Neurol 13: 227-231
2. Andersson M, Alvarez-Cermeño J, Bernardi G, Cogato I, Fredman P, Fredriksen J, Fredriksson S, Gallo P, Grimaldi LM, Grønning M, Keir G, Lamers K, Link H, Magalhães A, Massaro AR, Öhman S, Reiber H, Rönnbäck L, Schluep M, Schuller E, Sindic CJM, Thompson EJ, Trojano M, Wuster U (1994) The role of cerebrospinal fluid analysis in the diagnosis of multiple sclerosis: A consensus report. J Neurol Neurosurg Psychiatr 57: 897-902
3. Mehta PD (1991) Diagnostic usefulness of cerebrospinal fluid in multiple sclerosis. Crit Rev Clin Lab Sci 28: 223-251
4. Link H (1991) The cerebrospinal fluid in multiple sclerosis. In: Swash M, Oxbury J (eds) Clinical Neurology. Vol 2, Churchill Livingstone, Edinburgh, pp 1128-1139
5. Öhman S, Ernerudh J, Forsberg P, Roberg M, Vrethem M (1995) Lower values for immunoglobulin M in cerebrospinal fluid when sampled with an atraumatic Sprotte needle compared with conventional lumbar puncture. Ann Clin Biochem 32: 210-212
6. Thompson EJ, Keir G (1990) Laboratory investigation of cerebrospinal fluid proteins. Ann Clin Biochem 27: 425-435
7. Blennow K, Fredman P, Wallin A, Gottfries CG, Långström G, Svennerholm L (1993) Protein analyses in cerebrospinal fluid. Eur Neurol 33: 126-128
8. Watson MA, Scott MG (1995) Clinical utility of biochemical analysis of cerebrospinal fluid. Clin Chem 41: 343-360
9. Fieschi C, Gasperini C, Ristori G, Bastianello S, Girmenia F, Leuzzi V (1995) Patients with clinically definite multiple sclerosis, white matter abnormalities on MRI, and normal CSF: if not multiple sclerosis, what is it? J Neurol Neurosurg Psych 58: 255-256
10. Eilers RJ (1975) Quality assurance in health care: missions, goal, activities. Clin Chem 21: 1357-1367
11. Williams GW, Schork MA (1982) Basic statistics for quality control in the clinical laboratory. CRC Crit Rev Clin Lab Sci 17: 171-199
12. Reiber H (1995) External quality assessment in clinical neurochemistry: Survey of analysis for cerebrospinal fluid (CSF) proteins based on CSF/serum quotients. Clin Chem 41: 256-263
13. Olsson T, Kostulas V, Link H (1984) Improved detection of oligoclonal IgG in cerebrospinal fluid by isoelectric focusing in agarose, doubleantibody peroxidase labelling, and avidin-biotin amplification. Clin Chem 30: 1246-1249
14. Keir G, Luxton RW, Thompson EJ (1990) Isoelectric focusing of cerebrospinal fluid immunoglobulin G: an annotated update. Ann Clin Biochem 27: 436-443
15. Büttner J (ed) (1995) Proceedings of the IFCC meeting on reference materials and reference systems co-sponsored by WHO. Geneva, October 5-7 1994. Eur J Clin Chem Clin Biochem 33: 975-1022
16. Uldall A (1987) Quality assurance within clinical chemistry - a brief review emphasizing "good laboratory practice". Scand J Clin Lab Invest 47: 507-518
17. Miller JC, Miller JN (1989) Statistics for analytical chemistry. 2nd edn, Ellis Horwood Ltd, Chichester
18. Lefvert AK, Link H (1985) IgG production within the central nervous system: A critical review of proposed formulae. Ann Neurol 17: 13-20
19. Tibbling G, Link H, Öhman S (1977) Principles of albumin and IgG analyses in neurological disorders. I. Establishment of reference values. Scand J Clin Lab Invest 37: 385-390
20. McLean BN, Luxton RW, Thompson EJ (1990) A study of immunoglobulin G in the cerebrospinal fluid of 1007 patients with suspected neurological disease using isoelectric focusing and the log IgG-index. Brain 113: 1269-1289

Conclusions

P. Livrea

Since the thirties, the CSF examination has been the first laboratory test employed to support the diagnosis of MS [1]. In the following decades, the importance and the variety of the CSF abnormalities relevant for the diagnosis and follow up of the disease have become increasingly evident [2].

From sixties to eighties, the development and the wide use in large populations of patients of electrophoretic and immunochemical techniques for protein characterisation, employed alone or in combination, improved the assessment of sensitivity and specificity of the CSF abnormalities relevant for the disease diagnosis [3]. After the inclusion of the most common CSF abnormalities within a panel of standardised diagnostic criteria devised for MS research protocols, the investigation of CSF has received greater and greater attention, culminating in a "consensus" of experts which, at European level, clearly stated the parameters and the methodological procedures necessary to guarantee the quality of data essential for the management of MS at the diagnosis and during the disease course [4].

After the eighties and nineties, the further refinements of micromethods and the extension of the number of parameters measurable in CSF strongly stimulated the research for new, reliable and specific tests for the diagnosis or the follow up of the disease. This book summarises the most promising efforts performed in such a direction, but conclusions are that no CSF laboratory test has been yet validated neither for the diagnosis nor for monitoring MS activity.

The "historical" finding of the oligoclonal IgG increase in CSF of MS patients has been demonstrated to have high diagnostic sensitivity, but low specificity, moderate-low positive and high negative predictive values. The antiviral CSF antibody levels against measles, rubella and varicella zoster virus (MRZ reaction) seem to be a new parameter which should be included in the protocol of CSF investigations relevant for MS diagnosis [5], but its sensitivity, specificity, positive and negative predictive value must be still extensively studied. Other abnormalities, like the high frequency of an intra blood-brain barrier (BBB) IgM synthesis, the low frequency of intra BBB IgA synthesis, the high frequency of intra BBB synthesis of free kappa light chains [6], the predominance of IgG 1 and IgG 3 subclasses, or the presence of anti-MBP, anti-oligodendrocyte, anti-MAG, or anti-cerebellar soluble lectin antibodies, did not significantly improve the putative diagnostic value of a CSF profile specific for MS. Likewise, the fluctuations of CSF IgG oligoclonal patterns during the course of the disease, and/or their association with differently charged serum oligoclonal IgG components are not helpful neither for diagnostic purposes nor for following up the disease activity [7].

Crucial importance for the interpretation and the clinical use of CSF data stems from new concepts emerging in physiology and pathophysiology of CSF formation through a normal or a damaged BBB. The CSF flow in the different intra BBB compartments and the diffusion pathway dynamics which regulate the entrance of both solutes and cellular components from their sites of production into CSF, are now well known to play a paramount role in maintaining the steady state concentrations of these substances in CSF [8]. Differences in size, charge and receptor mediated mechanisms may change CSF/serum gradients of serum-derived CSF constituents; intracellular or intra CSF metabolic fate may affect the CSF/serum gradients of both serum-derived or intra BBB produced CSF components. Events which regulate the migration of leukocytes from blood into CNS, and from CNS to CSF spaces, are still poorly understood. A complex, likely specific, interaction between immunocompetent blood cells and cerebrovascular endothelia, including a variety of mechanisms like activation, recruitment, adhesion, deattachment, affect transendothelial migration from blood to CNS compartment; moreover, these mechanisms may differ for various leukocyte types, like polymorphonuclear or distinct mononuclear cell subsets [9].

New paradigms of blood CSF barrier function and dysfunction(s), and the knowledge of factors which regulate the turnover of cytokines, produced at site(s) of lesion(s) and acting in autocrine or paracrine fashion via fast interaction with high affinity receptors in target tissues, appear to be very important in understanding the biological significance of changes in CSF or serum levels of these substances. About this topic, the standardisation of methods, the quality of reagents, the sample timing and handling are all variables which require further reassessment [10].

MR imaging of brain and spinal cord has been shown to be a very sensitive and safe technique for the diagnosis and the follow up of the neuropathological activity in MS. The recent development of new imaging or spectroscopic techniques further improving the specificity of the abnormalities detectable by MR over the course of the disease, significantly changed the diagnostic approach to MS; MR imaging is now well accepted as a tool to monitor the activity of the disease and it has been considered a surrogate end point for evaluation of efficacy of new putative therapeutic procedures.

Taking into account the low specificity of CSF abnormalities and the invasive feature of a not easily repeatable procedure like the lumbar puncture, the MR imaging and spectroscopy seemed to bring about a further reduction of the laboratory role in the diagnosis and the follow up of a disease with an unpredictable clinical course like MS.

Few studies calculated the additional predictive value of CSF abnormalities obtained after the results of MR data in newly diagnosed MS patients [11]. In clinically probable MS patients diagnosed according to Shumacher criteria, MR imaging increased the confidence of a definite diagnosis in 50 % of cases, and did not confirm the presence of MS in 5% of cases; the results of CSF examination increased such percentages to 65 % in the case of definite MS, decreasing to 2% the ascertainment of a diagnosis different from MS; when clinically possible MS was considered, MR suggested a definite or a probable MS in 7% and in 23% of cases respectively, but did not confirm the diagnosis in 30% of cases. After CSF exam, the number of definite diagnosis increased to 15% and the number of patients not having MS increased to 35% of cases. If these data will be confirmed by a larger studies, also after MR imaging, 5-10% of not clinically definite MS patients may have a significant support for the diagnosis from CSF examination. The reduced frequency of a misdiagnosis or an uncertain diagnosis, even if in few cases, is obviously of great value in each individual patient, specially with the perspective of an effective therapy [12].

International criteria like those adopted for utilisation of CSF abnormalities in the MS diagnosis [13] have not assessed with comparable accuracy for other laboratory tests, which must be performed to exclude diagnosis different from MS in diseases presenting clinical and/or MR features mimicking MS. Some laboratory tests may be suggested in order to discriminate specific diseases, in selected clinical conditions: antinuclear antibodies, anti -DNA antibodies and other autoantibodies in serum; rheumatoid factor determination; complementoemia and immune complexes; human T-cell lymphotropic virus type 1 (HTLV-1) or human immunodeficiency virus (HIV) titres; Borrelia and Brucella serology; serum VDRL; Sjögren syndrome B and A; tests for tuberculous infection; angiotensin converting enzyme; serum vitamin B12 level; sedimentation rate or C-reactive protein; complete blood cell count; lactate/pyruvate or mitocondrial DNA analysis and muscle biopsy; very long chain fatty acids (VLCFA) in serum; ariysulfatase A in leucocytes [14].

However, comorbidity of MS with other disease, like specific infections, as supported by proper laboratory tests, cannot be excluded in some puzzling instances. Moreover, the presence of autoantibodies against a variety of organ and non organ specific antigens may occur in MS patients, independently from the presence of a systemic *lupus erythematosus* [15]. When comorbidity is unlikely and the specificity of the laboratory test is high, some laboratory data may display an unexpected usefulness. In a series of 120 patients with progressive spastic tetraparesis, gait ataxia, bladder dysfunction and paraesthesiae in the lower extremities, suggesting a primary progressive form of MS, displaying a normal MR of head and spine and lacking of oligoclonal IgG increase in CSF, we performed the assay of VLCFA levels in serum. In 4 cases the presence of an adult adrenomyeloneuropathy was demonstrated by high VLCFA levels. The clinical assessment and the finding of high serum VLCFA also in a number of the family members of the index cases confirmed the diagnosis.

In conclusion, the original application of CSF examination, specially when combined with other laboratory data, still maintains a basic, even if narrowed, role in the diagnosis of MS. New perspectives are becoming established for CSF studies also in the current MR imaging MS era. T2 weighted hyperintensities may represent different (co)presence of oedema, cellular infiltration, demyelination, gliosis, axonal loss; T1 gadolinium enhanced lesions may be represented by a reversible not lesioning BBB alteration or by a contemporary occurrence of cellular demyelinating infiltrate [16]. As changes in ICAM-1, lactate or cytokyne CSF levels seem to suggest, in the next future, other CSF parameters could be helpful to differentiate the various neuropathological characteristics of an apparently homogeneous lesion loading documented by MR imaging or spectroscopy. A well standardised, contemporary assessment of clinical, MR and CSF data will contribute to develop the "renaissance" of CSF studies in MS, which professor E.J. Thompson mentioned in the "Preface" of this book, likely increasing our capabilities in monitoring the natural history or the effects of treatments in MS neuropathology.

References

1. Merritt HH, Fremont-Smith F (1937) The cerebrospinal fluid. Saunders WB, Philadelphia
2. Lowenthal A, Van Sande M, Karcher D (1960) The differential diagnosis of Neurological diseases by fractioning electrophoretically the CSF γ-globulins. J Neurochem 6: 51-55
3. McLean BN, Luxton RW, Thompson EJ (1990) A study of immunoglobulin G in the cerebrospinal fluid of 1007 patients with suspected neurological diseases using isoelectricfocusing and the log IgG Index. Brain 113:1269-1289
4. Andersson M, Alvarez-Cermeno J, Bernardi G, Cogato I, Fredman P, Frederiksen J, Fredrikson S, Gallo P, Gronning M, Keir G, Lamers K, Link H, Magalhaes A, Massaro A, Ohman S, Reiber H, Ronnback H, Schluep N, Schuller E, Sindic CJM, Thompson EJ, Trojano M, Wurster U (1994) Clinical Neurochemistry for Diagnosis in Multiple Sclerosis: European Consensus. J Neurol Neurosurg Psychiatry 57 (8): 897-902
5. Felgenhauer K, Reiber H (1992) The diagnostic relevance of antibody specificity indices in multiple sclerosis and herpes virus induced diseases of the nervous system. Clin Invest 70: 28-37
6. Sindic CJM, Laterre EC (1991) Oligoclonal free kappa and lambda bands in the cerebrospinal fluid of patients with multiple sclerosis and other neurological diseases. J Neuroimmunol 33: 63-72
7. Zeman AZJ, Keir G, Luxton R, Thompson EJ (1996) Serum oligoclonal bands are a common and persistent feature of multiple sclerosis and have a systemic origin. J Med 89: 187-193
8. Reiber H (1995) Biophysics of protein diffusion from blood into CSF: The modulation by CSF flow rate. In: Greenwood J, Begley D, Segal M (eds) New concepts of a blood-brain barrier. Plenum Ldn, pp 219-227
9. Springer TA (1994) Traffic signals for lymphocyte recirculation and leukocyte emigration: the multiple step paradigm. Cell 76: 301-314
10. Gallo P, Piccinno MG, Tavolato B, Sidèn Å (1991) A longitudinal study on IL-2, sIL-2R, IL-4 and IFNγ in multiple sclerosis CSF and serum. J Neurol Sci 101: 227-232
11. Sharief MK, Thompson EJ (1991) The predictive value of intrathecal immunoglobulin synthesis and magnetic resonance imaging in acute isolated syndromes for the subsequent development of multiple sclerosis. Ann Neurol 29: 147-151
12. Giang DW, Grow VM, Mooney C, Mushlin AI, Goodman AD, Mattson DH, Schiffer RB and the Rochester-Toronto Magnetic Resonance Study Group (1994) Clinical diagnosis of multiple sclerosis. The impact of magnetic resonance imaging and ancillary tesing. Arch Neurol 51: 61-66
13. Poser CM, Paty DW, Scheinberg L, McDonald WI, Davis FA, Ebers GC, Johnson KP, Sibley WA, Silberberg DH, Tourtellotte WW (1983) New diagnostic criteria for multiple sclerosis: guidelines for research protocols. Ann Neurol 13 (3): 227-231
14. Goodkin DE, Doolittle DH, Hauser SS, Ransohoff RM, Roses AD, Rudick RA (1991) Diagnostic criteria for multiple sclerosis research involving multiply affected families. Arch Neurol 48: 805-807
15. Trojano M, Defazio G, Ricchiuti F, De Salvia R, Livrea P (1996). Serum IgG to brain microvascular endothelial cells in multiple sclerosis. J Neurol Sci (in press)

16. Filippi M, Miller DH (1996) Magnetic resonance imaging in the differential diagnosis and monitoring of the treatment of multiple sclerosis. Current Opinion in Neurology 9: 178-186

Subject index